Eight Easy Observing Projects

For amateur astronomers

David Cortner

and

Nancy L. Hendrickson

KALMBACH BOOKS

For my mom and dad, who believed the brightest stars were their own kids.—NLH

For my parents, who deserve my best book but get my first one instead.—DC

Printed in the United States of America

Book Design: Kristi Ludwig

Publisher's Cataloging in Publication
(Prepared by Quality Books, Inc.)

Cortner, David.
    Eight easy observing projects for amateur
astronomers / David Cortner and Nancy L. Hendrickson.
    p. cm.
    Includes bibliographical references and index.
    ISBN 0-913135-27-5

    1. Astronomy—Observers' manuals. 2. Astronomy—
Amateurs' manuals. I. Hendrickson, Nancy L.   II. Title.

QB63.C67 1996   523
                                QBI96-20220

Contents

Introduction

What do you do when the thrill is gone? Or when you first feel a nagging suspicion that there must be more to do with your new telescope, or with that telescope gathering dust in a closet since the last "event of the century" has come and gone?

Virtually any telescope represents a significant investment in money and time. A telescope is a durable symbol of your interest in seeing more of the world than most people ever imagine. We hope this book is filled with ideas and projects to help rekindle the enthusiasm that first motivated you to make that investment. If you're a beginner, we hope our book gets you (and your family) out into the night and looking up at the stars. If you're an old hand, we hope there are new perspectives and some surprising ideas here to enliven your astronomical life.

Have you purchased a telescope but been dismayed that you cannot see with it what you expected to see? This book will help you anticipate what you will really see when you go out under the stars. Precious few objects in the sky look as dramatic "in person" as they do in their photographs, no matter how large and how fine a telescope you use. Digital image processing has long been used to bring more information out of photographs, to sharpen what was unclear, to brighten what was dim. In this book, digital image processing is sometimes used to help narrow the gulf between how objects look in the eyepiece and how they appear on film. The difference between disappointment and wonder is often just a matter of knowing what to expect to see.

Maybe some aspect of your telescope works against your enthusiasm. If some critical part is not as good as it should be, a few dollars worth of hardware and an afternoon's work can make a rickety "department store telescope" into a pleasure to use. This book will help you choose accessories and replacement parts that will help get more from the instrument you already own, or help you select the one you ought to own.

Perhaps this book will help you show someone else why you spend so many hours out in the dark, or help you share your enthusiasm with a neighbor, a youngster, a classroom, or a club. It may remind you of other aspects of your hobby that you have not yet explored, or suggest ways to enrich your astronomical interest by involving others or by combining it with other hobbies, pastimes, and passions. Maybe you want to share your observations across time with future generations or just make the most of your life under the stars. Keeping a journal lets you do just that. We'll help get you started.

There are very few "observatory" photographs in this book. Except where noted they were made with off-the-shelf telescopes and accessories, often from a suburban backyard instead of world-class, back-of-the-beyond, black-sky sites. A chapter devoted to taking shapshots under the stars should help get you started if astrophotography is a project you want to make your own.

All this work, and at dawn only photographs and memories? Maybe not! A wide assortment of organizations need your expertise and your observations. This book will introduce you to some of the people and organizations who want your help keeping tabs on variable stars, storms on other planets, the nightly bombardment of meteors, surprising apparitions on the Moon, and the slow, decade-long heartbeat of the Sun. To help you find the books and products we've mentioned, see the Appendix on page 159.

Writing and illustrating this book has been an exciting project, too. We would never have started it and we could never have finished it without the encouragement and assistance of friends, relatives, and colleagues. Our thanks to our AstroForum and snail mail friends Steve Coe, Tom Polakis, Chuck Allen, Chuck Gulker, Ken Spencer, Gary Likert, Sheldon Cohen, Dan Ward, and Benoit Schillings. A very special thanks to Sissy Haas and Sue French, whose contributions to this book were invaluable; and to our editor, Terry Spohn, who was always at the other end of the lifeline. On a more personal note, thanks to George Kelley, Jr., Bill Harris, Harry Powell, Michael Castelaz, Ed Burke, Jr., Gary Henson, Preston Wolfe, Sheryl Garinger, Jessica Macbeth, Vicki Fite, Patricia Gleitsmann, and Charlene Crilley for their very different but always dependable support.

October 1995
Nancy L. Hendrickson
San Diego, California

David Cortner
Johnson City, Tennessee

1
Binocular Astronomy

For those of us who hit adolescence about the same time Sputnik hit the headlines, Space became a mania. We anxiously waited for long summer days to give way to evening. Streetlights flickered on and the world slowed down just enough to catch its breath. Suppertime came and went and then we were outdoors again, and when the blackness of Space touched the blackness of Earth, we looked up and saw the stars. Nothing in our 20th-century existence prepared us for the jolt. In that instant we had more in common with our Stone Age ancestors than we might have known, for once we had breathed starlight, the night sky owned part of our soul.

Whether you came of age during the days of Explorer and Pioneer, or Hubble and the shuttle, one thing is certain: Once the light from a galaxy over two million light years away strikes your eyes, you'll be hard pressed to settle for a naked-eye window on the universe. You may have dreamed of a telescope since you were a kid, or the bug may have left you unbitten until now, but there it is—you know you have to have a telescope. The only questions: Where to look, and what to buy.

The two major astronomy magazines, *Astronomy* and *Sky and Telescope,* are filled with telescope ads. Not only can you choose a refractor, a Schmidt-Cassegrain, or a reflector telescope, you also have your choice of German equatorial, fork, and Dobsonian mounts. Although it's tempting to just buy *something,* take the time to do some footwork. First stop: your local astronomy club.

There are hundreds of astronomy clubs in the United States (a complete list can be found in back issues of *Astronomy* or *Sky and Telescope* magazines), and in most of them you'll find other amateurs who'll go out of their way to help a newcomer. Go to a club meeting, then attend their monthly star party. Look through everyone's scope, compare models, designs, and costs, and ask lots of questions.

No astronomy club where you live? If you have a computer and modem, and belong to one of the major online services, you'll have access to a global network of other amateurs. CompuServe's Astronomy Forum (Go Astro), is among the finest. Its members include expert observers and astrophotographers who are knowledgeable in almost all astronomical subjects and are happy to answer your questions. The Astronomy Forum's Libraries also house files that include discussion threads on how to choose your first telescope. America Online's Astronomy Club also has discussion areas, as well as several downloadable astronomy software programs. Strangely, though, at both the local astronomy club and on the Astronomy Forum, an odd suggestion keeps cropping up. Instead of starting with a telescope, consider buying a pair of binoculars.

For under $150 you can buy a pair of binoculars that will let you see astronomical objects you never imagined. The Orion Nebula, Andromeda Galaxy, Jupiter's moons, countless clusters, *Mir,* and the Hubble Space Telescope are all easy binocular targets. In addition, learning the constellations is far easier with a star chart and a pair of 5-degree field-of-view binoculars than with a telescope with a ½-degree inverted field. And too, there are some objects that look far more spectacular in binoculars than a telescope. Binoculars are also the perfect instrument for satellite watching. Satellites move so fast it's difficult to find them in a telescope, and if captured, they're even harder to track. Using binoculars you can watch *Progress* bring supplies to the *Mir* space station, track the space shuttle and *Mir* as they play tag across the wide expanse of space, and pursue literally hundreds of weather, astronomical, and (sometimes) spy satellites.

Although binoculars won't show planetary detail, they can be used to track Jupiter's four largest moons as they perform their nightly ballet. And while lunar detail seen with a telescope can't be equaled with binoculars, with the help of even the simplest lunar map you can learn the major features and trace the extensive lunar ray systems during Full Moon. Even when properly filtered, though, binoculars aren't appropriate for solar astronomy.

Choosing binoculars

There are several factors to consider when picking the right pair of binoculars. First off, let's talk size. Binoculars are measured by their magnification and the diameter (aperture) of their main lens. For example, 10 x 50 means that objects appear 10 times closer (10 is the magnification) and the lens diameter is 50mm (25.4mm = 1 inch). Almost all have these numbers printed on them, and some also include the field of view, usually in degrees. Binoculars with greater magnification have smaller fields of view; so while the magnification of 10 x 50 is less than 20 x 80, you'll be able to see a larger area of the sky with the 10 x 50. Other factors to consider are exit pupil, eye relief, coating, type of prism, and focusing mechanisms. For a full discussion of those topics we turned to the editors of *Astronomy* magazine for their expert advice.

Fig. 1.1. The star cluster surrounding Alpha Persei, photographed rising through a cedar, is an easy target for binoculars. This photo mimics the cluster's appearance in 10 x 70 (or larger) binoculars. For help finding this sight, see the Cassiopeia starhop in Chapter 3.

Exit pupil

Aperture and magnification combine to produce a key specification for an astronomical binocular, the exit pupil. Divide the aperture by the magnification and you get the binocular's exit pupil. This is the diameter of the beam of light the binocular projects into your eye. For example, a 7 x 50 binocular has an exit pupil of 7mm (50 ÷ 7). A 10 x 50 binocular produces an exit pupil of 5mm. A giant 16 x 80 binocular also gives a 5mm exit pupil.

The best binoculars for astronomy are models with exit pupils from 5mm to 7mm. The diameter of the light cone projected from these binoculars nicely matches the diameter of the pupil of the human eye, assuming the eyes are dark-adapted to nighttime conditions. These binoculars are providing as much light as your eyes can accept and, as a result, are producing the brightest images for their aperture.

Most human pupils can open no wider than 7mm, so a binocular with an exit pupil wider than 7mm would waste light—only the center of the light cone could enter the eyes. In fact, there are no binoculars with exit pupils greater than 7mm. Those with exit pupils equal to 7mm (7 x 50, 8 x 56, 9 x 63, and 11 x 80 models, for example) are often called night glasses. These are some of the best models for astronomy.

As we get older, however, our pupils lose their ability to open wide at night. For those of us over 40, our pupils commonly open no wider than 5mm to 6mm, even when our eyes are fully dark-adapted. Under these conditions, binoculars with a 5mm to 6mm exit pupil are best. This includes 7 x 42, 8 x 42, 10 x 50, 15 x 80 and 16 x 80 models.

Eye relief

Eye relief is the distance your eye needs to be from the top of the eyepiece in order to see the entire field at once. As a general rule, wide-angle binoculars have

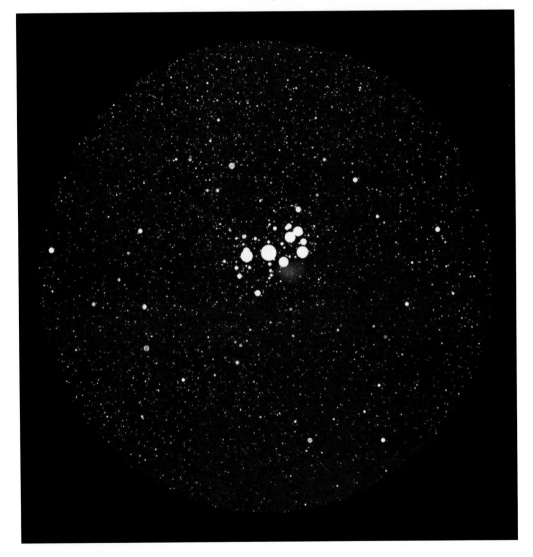

Fig. 1.2. Open cluster M45 (the Pleiades) as seen through a pair of 7 x 50 binoculars with a 7-degree field of view (limiting magnitude 10.5). The Merope Nebula is dimly visible.

reduced eye relief; your eyes need to be quite close to the eyepiece to see anything. An eye relief from 10mm to 15mm is normal for most eyepieces. Less than that and you have to press your eyes uncomfortably close to the eyepiece lenses.

If you wear glasses and wish to keep them on while viewing through your binoculars, you'll need binoculars with at least 15mm of eye relief. Models with eye relief of 20mm or more are ideal for eyeglass wearers. This requirement may rule out wide-angle binoculars for eyeglass wearers, as few of these models have eye relief greater than 15mm. The manufacturer's information, or the retail seller, should be able to provide you with eye-relief figures.

Coatings

Optical coatings placed on the surfaces of lenses and prisms increase light transmission (giving brighter images) and cut down on internal flare and reflection (increasing contrast). Most binoculars feature at least single-layer coatings on some air-to-glass surfaces. Better binoculars have all optical surfaces with one layer of coating and some surfaces treated with multiple coatings for even greater light transmission and contrast. In the best binoculars all optical surfaces are multicoated. Quite often, the difference in clarity between otherwise similar models of binoculars stems from the coatings.

When inspecting a pair of binoculars, look down into the main lenses. Do you see lots of white reflections from lights or windows? If so, it is a sign the optics are not multicoated or not coated at all. In a fully multicoated binocular, the reflections will be dim and will look deep green, blue, or purple.

Type of prism

Binoculars are compact because they use prisms to fold the light path back and forth within the body of the binocular. Porro prism binoculars are the most common and most affordable. These models have the familiar N-shaped bodies. Roof prism binoculars have straight tubes, making them more compact. But roof prism binoculars are more costly to make well, putting the best of these models in the premium price class.

For astronomy, all but the finest roof prism binoculars are outperformed by porro prism models. Roof prisms split the image into two and then recombine it, a process that can lead to spikes of light off bright stars on lower-cost models. Porro prism binoculars of any price tag don't have this quirk.

However, not all porro prism binoculars are equal. Ones with prisms made of BK7 glass are economical but suffer from some dimming of the edge of the field. Better binoculars use BaK4 glass prisms for the brightest field of view.

Type of focusing mechanism

Most binoculars focus with a central thumbwheel that moves both eyepieces back and forth in tandem. These "center-focus" models also have a separate adjustment on one of the eyepieces (usually the right eyepiece) to compensate for any difference in focus between your two eyes. Because focusing requires only one quick adjustment, center-focus binoculars are the easiest to use.

Some models use a focus adjustment on each eyepiece. Focusing these "individual-focus" models means always having to adjust two eyepieces. This is inconvenient and slow. The reason for going this route is that it is difficult to make center-focus binoculars waterproof. Individual-focus is the method of choice for marine binoculars. For applications such as astronomy, where the subject is always at the same distance, constant refocusing is not necessary, and the inconvenience of individual focus eyepieces is not a problem and may be a fair tradeoff for other features.

Recommended binoculars

For astronomy, everyone should own a pair from the 40 to 50mm aperture class. These are the lightweight, all-purpose binoculars you'll use most often, even after you acquire a telescope. Those wanting a true 7mm exit pupil or the longest eye relief should select a 7 x 50 or 8 x 56. Older viewers or those wanting slightly more power for greater resolution of small deep-sky objects should select a 10 x 50. Those wanting a slightly smaller, lighter binocular for serving double duty as a birding or daytime instrument should select a 7 x 42 or 8 x 42 model.

After you own a small binocular that can be held comfortably in your hands, you might want to consider buying a giant 70mm- or 80mm-class instrument. Essentially twin telescopes, these can provide stunning views of comets, deep sky object, and starfields. For a 7mm exit pupil and for the widest field, select a 10x or 11x model. For the higher power required to resolve many star clusters and to pick out small galaxies, select a 15x or 16x model. Big 20 x 80 binoculars are praised by some amateurs, but their narrow 3.5-degree field of view makes them more difficult to use for beginners. You also sacrifice eye relief when you move to higher power models.

Best binocular buys

Some of the best buys are found through mail-order sources. Astronomy magazines carry ads for stores like Orion Telescope Center and Pocono Mountain Optics, both of which carry astronomical binoculars. There's a wide selection of models and sizes, ranging from all-purpose 7 x 35 (commonly called birders) to giant 20 x 80 and beyond.

If you don't mind buying used binoculars, some great buys can be found in the pages of *The Starry Messenger,* a monthly publication made up solely of classified ads of used astronomy equipment. Those on the Internet can subscribe to an online listing of used astro equipment by sending an e-mail to listserv@netcom.com with the following in the body of the message: subscribe astro_fs. Finally, a sometimes forgotten source for used binoculars is your local camera store.

Binocular mounts

If you've chosen 10 x 50 binoculars, their light weight will let you hold them without a problem. However, after several minutes of holding even the lightest models, weight becomes a consideration. If you find your arms shaking, try propping the binoculars against a wall or post. For a extra support, keep your elbows tucked in close to your body.

You'll probably need to mount models larger than 10 x 50 on a photographic tripod, although some people can still hand-hold this size. Purchase an L-shaped binocular adapter from almost any telescope supply store and screw the bracket in the hole in the binocular's center post, then connect the other end of the bracket to the tripod head. However, as with a telescope, if your mount isn't steady, you'll hate it, so don't buy an ultra-lightweight tripod. If you can't find a hole in the center post, check for a little plastic button. On some models this unscrews to reveal the hole.

An alternative to the standard tripod head is the "quick-release plate" head. Instead of attaching the binocular adapter directly to the tripod attach it to a quick-release plate that snaps into place on the tripod. This way, when you're bino-cruising a star field and a satellite crosses your line of vision (and you will be amazed at how often this happens!) you can quickly release the binoculars and follow the satellite.

The major problem in using a tripod-mounted binocular is the difficulty in observing any object at the zenith (directly overhead). Several companies now market binocular mounts of varying design. Some are attached to your chair and some to your body; some include a chair; and some are parallelogram-shaped holders. Parallelogram mounts are designed so that no matter how many times the parallelogram is moved up or down to accommodate people of varying heights, the binoculars themselves are never moved away from the target.

Under the stars (at last!)

If the only thing you can pick out of the sky is the Big Dipper, purchase a planisphere or a simple star atlas, and a flashlight with a red lens covering. A planisphere allows you to "dial in" the date and time to see the currently visible night sky. In addition, every issue of *Astronomy* and *Sky and Telescope* contains a monthly all-sky chart. A good beginner atlas is the *Mag 6 Star Atlas*. It plots stars only down to the 6th magnitude (see insert), with the resultant chart being much less cluttered than an atlas which shows a deeper, and fainter, sky. Another popular atlas is *SkyChart 2000.0*. Why a red flashlight? It lets you see your charts without ruining your night vision.

If you'd rather plot your stars on the computer, there are several excellent planetarium programs. Although their features may vary, they all compute celestial objects seen from any location, at almost any date in time. Other features may include the ability to include or exclude objects (i.e., only display Messier objects, or stars brighter than magnitude 5, or just star clusters). Among the most popular software are the commercial programs *The Sky, Guide,* and *MegaStar* and the shareware package *SkyMap*.

When you're ready to tackle the sky, find as dark a site as possible; if it's

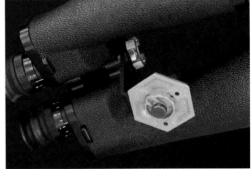

Fig. 1.3. 10 x 70 binoculars with an L-bracket used for mounting on any tripod head.

Fig. 1.4. Using a tripod quick release plate, mounted between the L-bracket and tripod head, allows fast mounting and fast removal of the binoculars from the tripod.

too light to observe from your own backyard, look for a darker site that's fairly close to home. The closer the better, because as many astronomers have discovered, if you have to drive an hour every time you want to observe, you probably won't keep it up.

For starters, learn the constellations, or at least the brightest ones. Some, like Scorpius the Scorpion and Orion the Hunter, look amazingly like their given names, and are quite easy to find. Others lie along dim stretches of the Galaxy, devoid of bright stars. These may be a challenge in city skies.

Once you can identify the constellations and some of the brighter stars (Vega, Sirius, Capella, Betelgeuse, Spica), find a bright star in the sky, locate it on your chart, then see if you can match up the visual star field in your

Fig. 1.5. Parallelogram-based binocular holders provide excellent support for heavy binoculars such as these 14 x 100s.

binoculars with the one on the chart. If you can, you've just taken the first step to finding your way around the sky.

Reading a star chart is a lot like reading a road map. Around the chart's edges are numbers which serve the same purpose that latitude and longitude lines do on Earth. If you know the celestial coordinates of an object, simply find those coordinates on your chart and then pinpoint your target.

Easy binocular targets

Clusters and associations. Galactic (or open) clusters are, without question, the most stunning of binocular targets. Composed of a few to over a thousand stars, the members share a common origin and travel through space bound by gravitational forces. While some of the fainter clusters appear in binoculars as faint clouds, many others have clearly definable members. Bright open clusters are the few objects that look about the same in your binoculars as their pictures do in astronomy magazines. Galactic clusters are quite young (astronomically speaking), with some still bathed in the wisps of interstellar gas and dust from which they were formed.

Another kind of cluster is the globular (or closed) cluster. Far, far older than galactic clusters, they are more compact, generally round in appearance, and can contain millions of stars. While galactic clusters contain blue or white stars (young stars), globulars are rich with red giants (older stars). Unfortunately for most people living in the United States, two of the brightest globulars (47 Tucanae and Omega Centauri) are too southerly to be seen, although Omega Centauri can be spotted about 10 degrees above the horizon in late spring from the very southernmost parts of the United States.

Through binoculars globulars most often appear as fuzzy balls, with individual members impossible to resolve. Among the easiest seen from the continental United States is M13 (in the constellation Hercules).

A third kind of star cluster is called an association or moving group. Like clusters, these groups move through space together, but not necessarily bound together by gravity. The best known asterism in the northern sky, the Big Dipper, is a member of the Ursa Major Moving Group.

Space birds. On most dark nights, if you watch long enough, you'll see satellites streaking across the sky. Shining brightly as they catch the Sun's rays from their high perch, they'll abruptly disappear as they move into shadow. What if you want to see a specific satellite—the space shuttle, for example —can it be done? Yes, if you have a computer, modem, and access to a bulletin board or online service that carries satellite orbital elements. The most common satellite prediction software is available through CompuServe, America Online, and the Internet. These include Sattrack (for UNIX), Orbitrack (Mac), SatSpy,

Traksat, and STSPlus (IBM PC). By entering your latitude and longitude, date and time, the programs compile a list of visible satellites observable from your location. What drives the software is something called orbital elements—the most current statistics about each satellite's orbit. Downloading the most recent elements every week or two will ensure that you have the latest positional information. To find satellite-prediction software and orbital elements:

CompuServe. Astronomy Forum (GO ASTRO) houses the major programs as well as elements. Search Library 3 for orbital elements (Bro key:elements), and Library 7 for prediction software (bro key:satellite).

America Online. Go to the Astronomy Club (keyword ASTRONOMY CLUB) and search the Astro Archives (prediction software) and General Files (orbital elements).

Internet. The Visual Satellite Observer's Home Page contains files on the basics of satellite observing, along with hypertext pointers to upcoming shuttle missions, the most recent *Mir* elements, satellite tracking software, and predictions of satellite passes for major United States and European cities. It also contains pointers to the Satellite Tracking Prediction Form (maintained by North Carolina State University) and Earth Satellite Ephemeris Service (Georgia State), two services that allow you to feed your own location data into the forms, which then create a list of visible sats.

VSO Home Page address:

http://www.physics.ox.ac.uk/sat/vsohp/satintro.html

To subscribe to the SeeSat-L mailing list for information on satellite observing techniques, software, re-entry predictions, and exceptional sightings, simply send a message with the word "subscribe" (without the quotes) in the subject field to:

Seesat-L-request@iris01.plasma.mpe-garching.mpg.de

Jupiter. For the binocular enthusiast, Jupiter appears as a bright disk with what appears to be four tiny stars moving in a circle around it. They are not stars, though, but four of Jupiter's 16 satellites. Io and Europa, the closest two moons visible through binoculars, are about the same size as Earth's Moon, while Ganymede and Callisto are larger. They orbit on about the same plane as Jupiter's equator, and appear as tiny pinpoints of light in a straight line on either side of the planet.

Moon. The map on page 20 lists the major lunar features, all visible through binoculars. See how many you can identify as the terminator (the line dividing night and day) moves across the Moon's surface. Note, too, how on the few nights following New Moon you can see the unlit portion of the Moon snuggle up to the thin lunar crescent. Called "the old Moon in the New Moon's arms," this phenomenon is attributable to "earthshine," the light reflected from Earth.

A binocular challenge

Sue French, a skilled observer and telescope maker, created a binocular observing program for her local astronomy club. Recipient of the Astronomical League's Messier Binocular and Binocular Deep Sky certificates, she began binocular observing in 1977. She is currently Vice-President of the Albany (New York) Area Amateur Astronomers, Inc., and on the board of directors. Sue's program includes easy as well as challenging binocular targets. To complete the project, you'll need a pair of binoculars (at least 7 x 50 recommended) and a star atlas or planetarium program. Good luck!

ALBANY AREA AMATEUR ASTRONOMERS

Binocular Observing Program

The Moon. The Moon is certainly the easiest object in this observing program to locate. The tricky part will be to sketch or describe the wealth of detail visible in even the smallest binoculars. Observe at any lunar phase and try to identify the most prominent features.

Jupiter. Any good pair of binoculars should reveal Jupiter's nonstellar nature, showing it as a tiny disk. They will also reveal the planet's four largest moons. Missing moons may be in transit, occultation, or eclipse. In addition, it is possible to lose one of the moons in the glare of Jupiter, especially if the binoculars do not come to a sharp focus. A sketch is the best way to show the positions of the moons.

Fig. 1.6. Two targets for the binocular observer, one easy, one a little tougher, are the brilliant Double Cluster and its fainter neighbor Stock 2 (near the top of this photo). Both are in in the constellation Perseus. The photograph spans about 4 degrees. Look for the Double Cluster with the help of the Cassiopeia starhop in Chapter 3.

Fig. 1.7. A NOAA weather satellite streaks across the North America Nebula. The North America Nebula is featured in the Cygnus starhop in Chapter 3.

M31[1] (Andromeda) 00h42.7m +41˚16'. At 2.2 million light years, the Andromeda Galaxy is the most distant object that can easily be seen with your eyes alone. Although the bright oval core may be all that meets the eye at first, be sure to look for the faint outer extensions of the spiral arms. Under very dark skies, the long dimension of the galaxy may be traced out for over 3 degrees.

M33 (Triangulum) 01h33.8m +30˚39'. The Pinwheel Galaxy is notoriously difficult for beginners with a telescope because it is very large and has a low surface brightness. Through binoculars you see the light concentrated in a smaller area with plenty of dark sky around it. M33 will appear about ½ degree across, the apparent diameter of the Full Moon.

Stock 2 (Cassiopeia) 2h15.0m +59˚16'. This large open cluster is composed of many faint stars spread out over an area of 1 degree. It fits in the field of view of most binoculars with the famous Double Cluster in Perseus. The brighter stars in this cluster can be imagined to form a stick-figure man.

NGC 869 (Perseus) 2h19.0m +57˚09'. This is the western member of the Double Cluster in Perseus, which is visible to the naked eye as a hazy patch in the sky. Binoculars will reveal two fairly bright stars that are surrounded by a swarm of fainter ones.

NGC 884 (Perseus) 2h22.4m +57˚07'. The eastern member of the Double Cluster. It

---

[1] "M" represents an object from the Messier catalog.

Fig. 1.8. Jupiter and its four largest moons. The moons, in order from left to right in both the binocular view and as seen by Voyager, are Ganymede, Io, Europa, and Callisto. Individual images courtesy NASA.

appears larger and more scattered than its nearby companion. Some of the stars in and around NGC 884 are distinctly red. See if you can detect any color.

Diamond Ring (Ursa Minor) 2h32m +89°. This asterism spans about ¾ degree and includes the North Star (Polaris). A dented circle of 7th-magnitude stars forms a ring with Polaris as the diamond.

Melotte 20 (Perseus) 3h22.0m +49°00'. This huge group of stars is better known as the Alpha Persei Association, a name it derives from its brightest member. Binoculars are the best way to view this group which spans over 3 degrees. It includes about 20 4th- to 7th-magnitude stars and dozens of fainter ones.

M45 (Taurus) 3h47.5m +24°07'. The Pleiades star cluster is a real binocular highlight. Naked-eye observers report anywhere from 6 to 20+ stars in this group, popularly known as the Seven Sisters. Binoculars will reveal dozens more. Once you have located some of the brighter stars with binoculars, see if you can use this knowledge to increase your naked-eye count.

Kemble 1 (Kemble's Cascade) (Camelopardalis) 3h58.0m +63°06'. This asterism consists of a 3-degree long chain of 8th- and 9th-magnitude stars running NW-SE. One 5th-magnitude star dominates this group. The SE end forks southward for about ½ degree in a straggling line and northeastward to a little cluster of stars (NGC 1502) that includes a pretty 7th-magnitude double.

Melotte 25 (Taurus) 4h27.0m +16°00'. Spanning over 5½degrees, the Hyades Cluster is best explored in binoculars. This group contains about 25 bright stars and many fainter ones. Several stars are bright enough to show color, and some nice wide pairs

Fig. 1.9. Map of some of the Moon's most prominent features, all easy targets for even 7 x 50 binoculars.

Craters

a Tycho
b Copernicus
c Plato
d Aristarchus
e Langrenus
f Grimaldi

Oceans & Seas

1 Sea of Rains
2 Serenity
3 Tranquility
4 Crises
5 Fecundity
6 Nectar
7 Clouds
8 Cognitum
   (the Known Sea)
9 Moisture
10 Ocean of Storms

may be spotted. Brilliant Aldebaran (α Tauri) is not a true member of the Hyades. Instead, it lies about halfway between us and the cluster. Aldebaran and the Hyades form the face of Taurus, the Bull.

M38 (Auriga) 5h28.7m +35˚50'. This open cluster is easy to spot in small binoculars, through which it looks like a small, hazy patch of light. Larger binoculars unveil a number of faint stars set amid an unresolved background. M38 lies within the distinctive pentagon of Auriga.

M42 (Orion) 5h35.5m -5˚23'. The Orion Nebula spans a full degree of sky. It is visible to the unaided eye, but binoculars greatly improve the view. Some of the hot, young stars that light this stellar birthplace can be seen.

Collinder 70 (Orion) 5h36m -1˚. This large, loose, open cluster surrounds and includes the stars forming Orion's belt. This group spans about 3 degrees and includes over 100 stars, most of which are visible in binoculars of 50mm or greater.

γ Leporis (Lepus) 5h44.5m -22˚27'. This is a nice binocular double with color contrast. As is true of many doubles, people frequently disagree about what those colors are.

M36 (Auriga) 5h36.3m +34˚08'. This open cluster is found in the pentagon of Auriga to the SW of M38. The two can fit in the same binocular field. M36 is smaller than M38, but it contains some brighter stars that are more easily resolved in binoculars.

M37 (Auriga) 5h53.0m +32˚33'. This open cluster is found just outside Auriga's pentagon. Only the central star is resolved in small to medium binoculars.

M35 (Gemini) 6h08.8m +24˚20'. In almost any binoculars, some stars may be resolved in this nice open cluster. The group is about ½ degree across and is located in the foot of Castor, one of the Gemini twins.

NGC 2244 (Monoceros) 6h32.4m +4˚52'. NGC 2244 is the open cluster at the heart of the Rosette Nebula. Some of the stars can be resolved in this cluster which is about ½ degree in diameter. Very dark skies may also reveal the faint 1.5-degree halo of the Rosette surrounding NGC 2244.

M41 (Canis Major) 6h47m -20˚46'. Binoculars will reveal many stars in this ½-degree open cluster, which can be seen as a hazy patch to the unaided eye on a dark transparent night.

M47 (Puppis) 7h36.6m -14˚29'. About 15 to 30 bright to faint stars may be counted in this loose, irregular, open cluster of about ½ degree.

M46 (Puppis) 7h41.8m -14˚49'. M46 lies just east of M47 in the same binocular field. It is very rich in faint stars that are only resolved in fairly large binoculars.

M48 (Hydra) 8h13.8m -5˚48'. Many moderate to faint stars can be seen through binoculars in this open cluster. In some of the older editions of *Norton's Star Atlas,* M48 will be found in the position marked with the Herschel designation 226, while the spot marked M48 is in error.

M44 (Cancer) 8h40.4m +19˚41'. At about 1.5 degrees, the Beehive, or Praesepe, Cluster is too large to fit in the field of most telescopes, but it is a nice binocular object. On clear, dark nights it is visible to the unaided eye as a misty patch, and binoculars can resolve many of its stars.

M81 (Ursa Major) 9h55.8m +69˚04'. This galaxy lies just behind the Big Bear's ear. In binoculars, M81 appears oval with dimensions of around 15 x 7 minutes of arc and shows much brightening towards the center. You may see the galaxy M82 in the same field. M82 looks smaller (8 x 2 minutes of arc), fainter and more elongated.

Melotte 111 (Coma Berenices) 12h25.0m +26˚00'. Melotte 111 is better known as the Coma Star Cluster and includes the stars that make up the dim constellation figure known as Coma Berenices. The brightest of its naked eye members lie a little over one third of the way from Denebola (β Leonis) to Alkaid (η Ursae Majoris). The entire cluster spans over 4.5 degrees.

γ Canum Venaticorum (Canes Venatici) 12h45.1m +45˚26'. γ Cvn is a variable star, but it is included in this observing program because of its color. This star, sometimes referred to as La Superba, is one of the reddest stars in the sky. Hovering around the limits of naked-eye visibility, look for it about one-third of the way from Cor Caroli (α Canum Venaticorum) to Megrez (8 Ursae Majoris).

M51 (Canes Venatici) 13h29.9m +47˚12'. The Whirlpool Galaxy can be seen in most binoculars as a faint round glow with a small, brighter nucleus. Large binoculars will reveal the small companion galaxy at its northern edge.

M3 (Canes Venatici) 13h42.2M +28˚23'. This globular cluster resides in a lonely looking area of the sky. Look for a fuzzy, round patch of light about halfway along a line from Arcturus (α Bootis) to Cor Caroli (α Canum Venaticorum).

M4 (Scorpius) 16h23.6m -26˚31'. M4 is fairly large for a globular cluster. Visually it spans about 12 minutes of arc and shares the field with brilliant Antares (α Scorpii).

M13 (Hercules) 16h41.7m +36˚28'. The Great Globular Star Cluster in Hercules is faintly visible to the unaided eye under dark, transparent skies. Binoculars will easily show this bright globular as a round glow in the Keystone of Hercules.

M92 (Hercules) 17h17.1m +43°28'. This nice globular cluster is often overlooked by observers intent on its more famous neighbor, M13. Through binoculars, M92 brightens considerably toward the center which shows a nearly stellar-looking nucleus.

υ Draconis (Draco) 17h32.2m +55°11'. This double star is the faintest of the four stars that form Draco's head. It is an easy binocular split with a separation of 62".

M6 (Scorpius) 17h40.0m -32°12'. The Butterfly Cluster shows about five fairly bright stars and a number of fainter ones through binoculars. In low-power binoculars, it shares the same field with the larger and brighter cluster M7.

IC 4665 (Ophiuchus) 17h46.3m +5°43'. This is a very large, loose open cluster of about a dozen bright stars mixed among fainter members. It is over a degree in diameter.

M7 (Scorpius) 17h54.0m -34°49'. M7 is a beautiful binocular cluster almost a degree across. Binoculars reveal many member stars, which often twinkle wildly because of their low altitude in our sky.

Fig. 1.10. Kemble's Cascade high in the northern sky in Camelopardalis, west of Cassiopeia. The photo spans about 4 degrees. This cluster is included on the Cassiopeia starhop in Chapter 3.

M8 (Scorpius) 18h03.7m -24°23'. NGC 6523 is the Lagoon Nebula, while NGC 6530 is its embedded star cluster. Together they form a naked-eye object over the spout of the Teapot. Binoculars disclose a sprinkling of mixed bright and faint stars surrounded by a misty glow.

M24 (Sagittarius) 18h18.4m -18°25'. The Small Sagittarius Star Cloud is a detached section of the Milky Way spanning 120 x 40 minutes of arc. It is sometimes incorrectly identified as NGC 6603, a small open cluster embedded in M24.

M22 (Sagittarius) 18h36.4m -23°54'. This globular cluster rivals M13 in brightness, beauty and size, but it rides much lower in our sky. It is easily visible in binoculars as a round, hazy glow.

IC 4756 (Serpens Cauda) 18h39.0m +5°27'. IC 4756 is a large open cluster covering about 1 degree of sky. It is rich in faint to very faint objects.

ε Lyrae (Lyra) 18h44.3m +39°40'. Epsilon Lyrae is the famous Double-Double star. Binoculars clearly show ε1 and ε2 widely split. Some people can even detect their dual nature with the unaided eye. The closer components of ε1 and ε2 are telescopic objects.

Collinder 399 (Vulpecula) 19h25.4m +20°11'. This group is also known as Brocchi's Cluster or the Coathanger. It includes the naked-eye stars 4, 5 and 7 Vulpeculae, and spans a full degree. The cluster resembles the old-fashioned coathanger with a curved wooden bar and metal hook.

M27 (Vulpecula) 19h59.6m +22°43'. This planetary nebula is commonly known as the Dumbbell or Apple Core Nebula. Through small binoculars M27 may look rectangular. Larger ones show the apple core shape and may reveal the faint outer extensions that round it out.

α Capricorni (Capricornus) 20h18.1m -12°33'. α Capricorni is an optical double whose components are separated by over 6 minutes of arc. Many people can see this with the unaided eye, while binoculars show a wide split. Both components have a yellow hue.

β Capricorni (Capricornus) 20h21.0m -14°47'. Beta Capricorni is another easy binocular double and makes a pretty sight in the same field with Alpha Capricorni. Some subtle color contrast may be noticed.

NGC 7000 (Cygnus) 21h01.8m +44°12'. At about 120 x 100 minutes of arc, the North American Nebula is too large to fit in the field of view of many telescopes, but binoculars can encompass the full breadth. While some observers may not be able to discern the North America shape, the dark clouds that define the outer edges of NGC 7000 are quite obvious.

M15 (Pegasus) 21h30.0m +12°10'. This globular cluster is an easy binocular target found about 4 degrees NW of Enif (ε Pegasi), the Horse's nose. It shows considerable brightening toward the center and a bright, nearly stellar core.

NGC 7293 (Aquarius) 22h29.7m -20°47'. This planetary nebula is more well known as the Helix. It has a very low surface brightness and can be difficult to spot through a telescope. Binoculars concentrate the light of this ¼-degree object into a smaller area, making it easier to pick out.

NGC 7789 (Cassiopeia) 23h57.0m +56°44'. This open cluster consists of many extremely faint stars. Small binoculars show a fairly large hazy patch about 16 arc minutes across. Large binoculars reveal many faint pinpoints of light against an unresolved background.

Fig. 1.11. Brocchi's Cluster, the Coathanger. Conspicuous in any binoculars, in southernmost Cygnus. The field of view is about 4 degrees across. To find this cluster, see the Cygnus starhop in Chapter 3.

2

Your First Telescope
The 60mm Refractor

It's red, it's shiny, it has lots of accessories, and it's yours. The box says 500 power(!), and you can't wait to crank it to the max. It's your first honest-to-goodness telescope, and whether it appeared under the Christmas tree when you were ten, or was purchased on an impulse when you were 40 doesn't matter—this is the scope you'll never forget.

If you're like most, you'll probably spend the morning handling all the components, poring over the manual, then lining up the finderscope on a TV antenna three blocks down the street. Finally, in late afternoon, you'll carry the scope out to the backyard, waiting for night. If you're lucky the sky will be

Fig. 2.1. A commonly chosen "first scope," a 60mm refractor, remounted on a homemade pipe mount.

clear, steady and dark. Finally comes "first light"—that moment when starlight ends its long journey through the vastness of time by striking your glass, your eye, and then your heart. For some, this becomes a ritual played out with each successive telescope, the searching out of a star or planet held dear (a personal favorite or a historical beacon) to initiate new glass. From the smallest spyglass to the great observatories, first light is special light.

You check your star maps. What will be easy to find, situated in a clear chunk of sky unobstructed by trees, houses, or streetlights, and be memorable? One of summer's guideposts, Altair, Vega, or Deneb? Or wintery Orion's ruddy Betelgeuse or blazing blue-white Rigel? Whatever you choose, it will remain special.

For many, first light comes from the Moon or planets. The Moon is especially hard to resist 'cause it's big, bright, and easy to find. Even at low power the lunar surface beckons you to explore its valleys and peaks, domes and rilles. With the naked eye the crisp line dividing night and day seems impenetrable, but at the eyepiece, mountains grounded in darkness thrust high enough above the surface for their peaks to catch the dawn's rays.

No matter what the target (or the instrument), the results are the same. Like Galileo and Messier, Herschel and Hubble, you've looked through the eyepiece and made the night sky your own.

However, on successive clear nights, as you walk the familiar path to your scope, troublesome thoughts cross your mind. "Why am I having trouble finding objects in the finderscope? Why does the scope shimmy like it was caught in a high wind? How come I can't see anything with my high-power eyepiece and Barlow? Am I doing something wrong?"

If you're asking yourself these questions, you're on the right track.

The optics of a 60mm (2.4-inch) refractor generally perform well on bright objects. However (and this is a big however), the 60mm comes equipped with accessories that make it hard to use! Once more: all those nifty accessories, including the undersized finderscope, the lightweight tripod and mount, and the small diameter eyepieces are a hindrance to easy observing. Here's why.

The lucky fellow who owns a several-thousand-dollar refractor has both the highest quality glass, and a mount like the Rock of Gibraltar. He can touch it, bump it, walk all around it, and it will not move! You, on the other hand, may spend more time on post-focus jiggle abatement than you do on stargazing. Also, your finderscope has a narrow field of view; in no way does its dinky slice of sky resemble what you see on the chart. And, a .965-inch-diameter eyepiece is no match for a 1¼-inch eyepiece. Period.

Finally, the higher the magnification, the narrower your field of view, and the quicker the object will zip across the eyepiece. A 2x Barlow working in conjunction with a 4mm eyepiece will give the ultimate in magnification and the least enjoyable viewing. The good news is there are easy fixes.

Most important, the mount and tripod need weight and stability. There are a couple of approaches to the problem. To beef up the tripod, take a trip to a hardware store or junkyard and dig out a hunk of metal weighing 10 or 15 pounds. Either suspend the weight between the tripod legs or set it on the metal eyepiece tray. If you don't have a piece of metal, a bag of sand will work too. With the added weight you'll see an immediate improvement. Next, try throwing rubber casters under the tripod feet—they'll dampen the vibrations when anyone walks by. The goal is to create as rock-steady a tripod as you can. Nothing will put you off astronomy more quickly than a bad case of the shakes.

Fig. 2.2. A 5-inch apochromatic refractor built by Roland Christen of Astro-Physics riding a mount built by Scott Losmandy to hold an 11-inch Celestron Schmidt-Cassegrain. A 3-inch Unitron refractor is mounted as a guide telescope. It takes a determined bump to accidentally move this instrument! (This outfit was used to make many of the photographs in this book—electric focusing by Jim's Mobile, film hypered by Jack Marling's Lumicon, and CCD-based tracking by SBIG complete the credits.)

You could stop the upgrades right here and probably be happy, but if you want to go further, changing the mount itself is the next step. It's entails a little more work, but it's well worth the results. What you're going to do is replace the high-tech appearing equatorial mount with a "pipe mount"—an unattractive low-tech substitute that's simplicity in action. Here's what you'll need:

Two 1-inch metal floor flanges
Assorted screws
One 6-inch length of pipe (1 inch in diameter)
One 90-degree elbow joint (1 inch in diameter)
Pair of muffler clamps (2.4 inches in diameter—to fit your scope)
Lapping compound (plumbing or auto supply store)
Scrap wood

The only part of the assembly that takes anything more than household tools is crafting a block of wood to fit between the tripod legs to serve as a base for the new mount. Fortunately you'll only need a jigsaw (or a generous neighbor) and minimal skill. When you finish cutting the block of wood to fit between the tripod legs, sand off the rough edges, attach it to the tripod, and you're in business. Before you go further, put lapping compound on all of the metal threads. Screw and unscrew the fittings until the action is smooth.

Next, using screws, attach one of the flanges to the wooden "mount" block. Into the flange thread the 6-inch length of pipe. Next comes the 90-degree elbow joint.

Fig. 2.3. Craft a wooden head from scrap lumber using a jigsaw, then mount it between the standard tripod legs.

Fig. 2.4. Attach a 1-inch metal flange to the wooden head.

Fig. 2.5. All of the components needed to replace a 60mm refractor's standard, lightweight equatorial mount with a hefty steel pipe mounting. Carefully prepared threads provide smooth bearing surfaces.

The only thing left is to attach the scope to a cradle, and the cradle to the mount. Use a flat piece of wood as the telescope's cradle (the part the scope will rest on). Attach the scope to the cradle using a pair of muffler clamps. The 2.4-inch size fits snugly around the scope, holding it firmly. If your scope moves around, either buy smaller clamps or pad the clamps. You're looking for a snug fit. On the bottom of the cradle, attach the second flange. This flange screws onto the 90-degree elbow joint. Your pipe mount is finished!

Once the tripod and mount are beefed up, take off the original finderscope and replace it with a Telrad®, a squared-off, bulky-looking finder with no magnification. When you look through it, you'll see three concentric rings projected onto your field of view, representing circles of ½ degree, 2 degrees, and 4 degrees in diameter. With the large field of view, you'll find it easier to line up on naked-eye stars. Ads for Telrads® are in the popular astronomy magazines.

The next fix is one you may or may not choose to make. If you want to use larger-diameter eyepieces, you can convert your eyepiece holder from .965-inch to one that accepts eyepieces 1¼ inches in diameter by using a "hybrid diagonal." Hybrids can be purchased from stores or mail order houses carrying astronomical equipment. As an alternative, consider replacing the inexpensive .965-inch eyepieces of Ramsden or Huygenian design with ones of Orthoscopic design.

As for the high magnification: Although the manual that came with your scope may tout using a Barlow with your shortest-focal-length eyepiece for great magnification, don't do it! A short-focal-length eyepiece combined with a 2x Barlow will furnish an *extremely* narrow field of view. If you want to use the Barlow, use it with your longest-focal-length eyepiece (generally a 20mm). This way you'll have a second eyepiece of 10mm, *and* retain the longer eye relief of the 20mm one.

Favorite targets for a 60mm refractor

The Moon. The Moon is such a popular target we saved a whole chapter for it, but until later, here's a little to get you going. For any observer, beginning or not, the Moon is a proving ground for sharpening observing skills. As in most endeavors, practice makes perfect. This means that the more frequently you observe an object, the more you will see. Sound impossible? Give it a whirl—there's just no substitute for eyepiece time. Does this mean you won't see much the first time out? Not at all. In fact, the lunar surface will probably overwhelm you with its sheer number of features.

A good lunar atlas or map, such a Antonín Rükl's *Atlas of the Moon* (Kalmbach Publishing), Phillips' *Moon Map*, or Hallway's *The Moon*, will help

Magnification

Just how powerful is your telescope? Power (or magnification) is determined by the eyepiece you're using. To determine magnification, simply divide the focal length of the telescope by the focal length of the eyepiece. For example, a 10mm eyepiece used on a 1200mm telescope will yield 120x; a 20mm eyepiece will give 60x. Although 120x doesn't seem much compared to the manufacturer's claims of 500x, it's actually the largest *useful* magnification for a 60mm (2.4-inch) scope. The highest useful magnification of any telescope is considered to be 50 or 60x per inch of aperture (or 2x per millimeter).

Fig. 2.7. Attach the telescope to the wooden cradle using muffler clamps.

catalogued several thousand pairs, and American amateur S. W. Burnham discovered 1200 doubles over the course of his lifetime.

Over the last 15 to 20 years the interest in doubles has waned, partly because as amateur instruments grew bigger and better, the ability to observe deep-sky objects increased. The quest for devouring the "faint fuzzies" with 12- to 20-inch backyard scopes has supplanted the love of double star observing done by our 19th-century counterparts with their 3-inch refractors.

Why observe doubles? It's a realm of astronomy perfectly suited for a small refractor operating in light-polluted skies. While faint galaxies, with low surface brightness, are invisible in urban lights, a double star doesn't suffer the same fate.

Depending on the magnitude[1] of each of the component stars, and the distance they are from each other, some of them can be resolved, or "split"—which means when you look through your telescope at a seemingly solitary star, you can see one or more traveling companions. If you can't clearly resolve them, they may be too close together, or the components' magnitudes

---

[1] See Chapter 3 for completed discussion of magnitude.

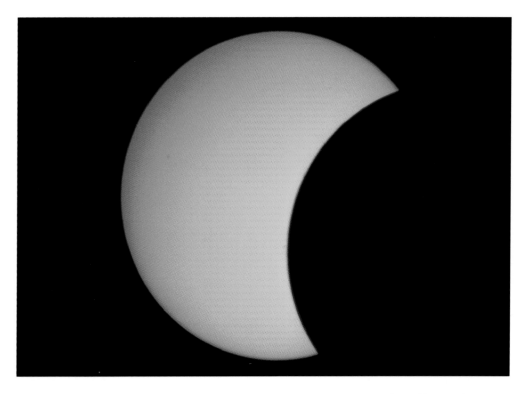

Fig. 2.8. The Sun, with its ever-changing sunspots, is an excellent target for a 60mm refractor appropriately fitted with a full-aperture solar filter, as are solar eclipses: Not long till totality—and the sky is still clear!

may vary so greatly that the glare from the brighter of the pair wipes out the dimmer one. As a rule, a pair equal in magnitude is easier to resolve that an unequal one, and brighter pairs are easier to resolve than dimmer ones.

Although recent astronomy books have, for the most part, ignored double stars, you can create your own observing list from a star chart, planetarium software, or on a constellation-by-constellation search through Burnham's *Celestial Handbook*. A search through a used book store can yield older, out-of-print books that list favorite doubles. These books give the magnitudes of the stars, as well as their amount of separation. Separation is measured in arc seconds (1 arc second equals ⅟3600 degree). Older books are often advertised in *The Starry Messenger*. Astro Cards, a Pennsylvania company, publishes a set of 3 x 5 cards called *The Finest Double Stars*. The Astro Cards picture each constellation, marking the positions of the doubles. They also chart the -magnitudes and positions of the doubles, along with any special notations of star color.

When you first start splitting doubles, begin with the easy ones: Riccioli's Mizar, Albireo (in Cygnus), or Eta Cassiopeiae. Begin with pairs that have at

Continued on page 40

Sissy Haas's Year-Round Double Star Observing Program

SPRING	R.A.	Dec.	Magnitude	Sep. (in " of arc)	Power	Comments
Chi Bootis	14h13.5	+51° 47'	4½/6½	13.4	25	Pure white with gray. Forms double-double with Iota
Iota Bootis	14h16.2	+51 22	5/7½	38.5	25	Pure white with distant colorless, in field with Chi
Pi Bootis	14h40.7	+16 25	5/6	5.6	80	Snow white, easy at high power
Epsilon Bootis	14h45.0	+27 04	2½/5	2.8	120	Glorious but difficult. Brilliant orange
Mu Bootis	15h24.5	+37 23	4½/6½	108.3	25	Bright, superwide, yellow-white
Zeta Cancri (A-C)	8h12	+17 39	5½/6½	5.7	45	Pretty, pure yellow
Iota Cancri	8h46.7	+28 46	4½/6½	30.5	25	Bright, easy, with deep color contrast. Sun yellow/ocean blue
24 Cancri	8h26.7	+24 32	7/8	5.8	45	Semi-faint, but easy at medium power, orange-yellow
24 Comae Berenices	12h35.1	+18 23	5/6½	20.3	25	Orange with blue, easy
16/17 Draconis	16h36.2	+52 55	5/5½	90.3	25	Bright wide, very pretty, pure white
Mu Draconis	17h05.3	+54 28	6/6	1.9	120	Semi-faint "peanut" at high power
Nu Draconis	17h32.2	+55 11	5/5	61.9	25	Webb called "grand object." Seems like pair of headlights.
Omicron Draconis	18h51.2	+59 23	4½/7½	34.2	25	Great contrast, but not difficult. Yellow-orange with dim gray
Alpha Leonis	10h08.4	+11 58	1½/8	176.9	25	Brilliant bluish white
Gamma Leonis	10h20.0	+19 51	2½/3½	4.4	120	Brilliant touching, yellow-orange, use highest power
54 Leonis	10h55.6	+24 45	4½/6½	6.5	45	Nice at medium power, white with touching gray
93 Leonis	11h48.0	+20 13	4½/8½	74.3	25	Dramatic contrast, brilliant yellow
Alpha Ursae Majoris	11h03.7	+61 45	2/7	378	25	Brilliant yellow-white with distant colorless, easy
Zeta Ursae Majoris	13h23.9	+54 56	2½/4	14.4	25	Brilliant, easy pair at easiest location. In field with Alcor
Alpha Ursae Minoris	2hr31.8	+89 16	2/9	18.4	45	Glorious but extremely difficult, brilliant yellow—orange
Gamma Virginis	12h41.7	-1 27	3½/3½	1.8	120	Brilliant rod to peanut at high power. Yellow-white
54 Virginis	13h13.4	-18 50	6/7	5.4	45	Best at medium power, but touching. Semi-faint.

SUMMER	R.A.	Dec.	Magnitude	Sep.	Power	Comments
5 Aquilae	18h46.5	-0 58	5½/7½	13.0	25	Pure white with gray. Low power best.
15 Aquilae	19h05.0	-4 02	5½/7	38.4	25	Wide, easy, beautiful. Pure gold with blue-green
Beta Cygni	19h30.7	+27 58	3/5	34.4	25	Beautiful contrast, favorite of many, brilliant yellow-orange/deep blue
16 Cygni	19h41.8	+50 32	6/6		25	Wide, easy, in rich field. Orange twins.
Omicron Cygni	20h13.6	+46 44	4/5/7	107-337	25	Easy triple. Orange superwide with closer blue
61 Cygni	21h06.9	+38 45	5½/6	29.5	25	Famous twins with possible planet. Only 11 light-years away. Orange
Gamma Delphini	20h46.7	+16 07	4½/5½	9.6	45	Glorious at medium power, brilliant orange-red, close
Chi Herculis	16h08.1	+17 03	5/6	28.2	25	Very nice, unusual orange with red in rich field
36/37 Herculis	16h40.6	+4 13	6/6½	69.8	25	Wide and easy, seems brighter, yellow-white
Alpha Herculis	17h14.6	+14 23	3/5½	4.7	90	Dramatic, colorful contrast, barely separated. Red/orange-blue/green
95 Herculis	18h01.5	+21 36	5/5	6.3	45	Nice at medium power, perfect twins, touching, orange-yellow
Epsilon Lyrae	18h44.3	+39 40	5/5 5/6	2.6/2.7	120	The "double-double." Two tight pairs in same field, difficult
Beta Lyrae	18h50.1	+33 22	3½/7	45.7	25	White with wide blue in rich field
54 Sagittarii	19h40.7	-16 18	5½/8½	45.6	25	Yellow-orange with distant colorless
Beta Scorpii	16h05.4	-19 48	2½/9	13.6	25	Bright, easy, nice contrast. Brilliant yellow-white with blue

35

FALL	R.A.	Dec.	Magnitude	Sep.	Power	Comments
Gamma Andromedae	2h03.9	+42 20	2½/5½	9.8	45	Glorious contrast, very easy. Brilliant orange, very deep blue
59 Andromedae	2h10.9	+39 02	6/6½	16.6	25	White near twins, easy but semi-faint
Zeta Aquarii	22h28.8	-0 01	4½/4½	2.1	120	Brilliant "peanut" at high power
94 Aquarii	23h19.1	-13 28	5/7	12.7	25	Nice contrast. Yellow-white
107 Aquarii	23h46.0	-18 41	5½/6½	6.6	45	Fine at medium power, but touching. Pure white
Gamma Arietis	1h53.5	+19 18	4½/4½	7.8	35	The ram's eyes. Brilliant twins barely separated, snow white
Lambda Arietis	1h57.9	+23 36	5/7½	37.4	25	Nice colors, yellow with blue, wide and easy
30 Arietis	2h37.0	+24 39	6½/7½	38.6	25	Wide and easy, but semi-faint. White to colorless.
Alpha Capricorni	20h17.6	-12 30	3½/4	377.7	25	Brilliant superwide, near twins. Pure yellow
Beta Capricorni	20h21.0	-14 47	3/6	205.3	25	Brilliant superwide with color contrast, orange with blue
Omicron Capricorni	20h29.9	-18 35	6/6½	21.9	25	Easy near twins in low sky. Mainly gold
Chi Cephei	20h08.9	+77 43	4½/8½	7.4	25	Nice but difficult. Yellow-orange, companion barely seen
Beta Cephei	21h28.7	+70 34	3/8	13.3	25	Striking contrast, brilliant snow white
Mu Cephei	21h43.5	+58 47	4		25	Single star but noteworthy. Extremely deep orange, the "garnet" star
Epsilon Equulei	20h59.1	+4 18	5½/7	10.7	25	Nice contrast but difficult location. White with blue-gray

WINTER	R.A.	Dec.	Magnitude	Sep.	Power	Comments
41 Aurigae	6h11.6	+48 43	5/7	7.7	45	Both white, seem more equal
32 Eridani	3h54.3	-2 57	5/6	6.9	45	Very nice at medium power, but barely separated Orange and blue
20 Geminorum	6h32.3	+17 47	6/7	20.0	25	Orange with gray. Easy
38 Geminorum	6h54.6	+13 11	4½/7½	7.1	45	Very nice at medium power, white and touching dim gray
Alpha Geminorum	7h34.6	+31 53	2/3	3.5	90	Brilliant touchers, easy at high power, yellow-white
Gamma Leporis	5h44.5	-22 27	4/6	94.6	25	Deep unusual colors, yellow with red, requires clear night
Epsilon Monocerotis	6h23.8	+4 36	4½/6½	13.4	25	Nice contrast, yellow with gray, easy
Beta Monocerotis	6h28.8	-7 02	5/5½/6	7.3/10	45	Fantastic triple, all three identical, tight pair with wider third
23 Orionis	5h22.8	+3 35	5/7	32.1	25	Snow white with wide gray, easy
Lambda Orionis	5h35.1	+9 56	4/6	4.4	90	Beautiful contrast, but a little difficult. Orange with blue
Theta Orionis	5h35.3	-5 23	7/8 (AB) 7/5 (AC) 7/7 (AD)	8.8 12.8 21.5	90 90 90	Tight fantastic quadruple inside the great nebula. A little difficult, especially member B. All snow white
Iota Orionis	5h35.4	-5 55	3/7	11.4	45	Profound contrast, but not too difficult. Brilliant orange, dim blue
Struve 747	5h35.0	-6 00	5½/6½	35.7	45	Easy bonus in field with Iota, pure white, wide
Eta Persei	2h50.7	+55 54	4/8	28.3	25	Dramatic contrast, but wide and easy. Yellow-orange/deep blue
62 Tauri	4h24.0	+24 18	6/8½	28.9	45	Nice contrast but difficult.
Theta Tauri	4h28.7	+15 52	3½/4	337.4	25	Beautiful twins within Hyades, superwide, pure orange
118 Tauri	5h29.3	+25 09	6/6½	4.8	45	Fine at medium power, orange with touching blue

least 8 to 10 seconds of arc separation. On closer doubles, using averted vision (glancing out of the sides of your eyes) may help, as the sides of the retina are more sensitive than the center.

Without question, the most exquisite of doubles are those of contrasting color—Albireo's yellow and blue, Eta Cassiopeiae's red and yellow, Regulus' emerald and crimson. For once, what you see with your own eyes, through your own scope is *far* superior to any photograph of a multiple star system. Although the colors we see are sometimes due to the star's spectral type, most often we are seeing a color contrast because a star takes on the complementary color of a brighter star next to it.

In the 19th century, our astronomy colleagues had a passion for naming the colors of double stars. Although you might think red is red and yellow is yellow, they did not. In his delightful book, *Soul of the Night,* Chet Raymo notes that 19th-century observers saw colors "more suited to the garden than the sky." These included straw, rose, grape, and lilac. Raymo went on to note that famed double star observer William Henry Smythe had an eye "apparently refined enough to see a dozen shades of white including pearly, lucid, cream, silvery and just plain whitely white." And if that's hard to believe, twist your tongue around Wilhelm Struve's star color "olivaceasub-rubicunda," or pinkish-olive! Are there stars that are topaz yellow, apple green, or flushed white? Only you can decide.

A year-round double star observing program

Double star observers don't come any better than Pennsylvania amateur Sissy Haas. Observing from a backyard light enough to read star charts, Sissy found that if she counted on trips to dark sky sites, the amount of time she spent observing lessened as the hassle of packing up equipment increased. She decided to save her bigger reflector for the occasional dark sky trips, and instead returned to the 60mm refractor that many consider "junk." Instead of observing the "faint fuzzies," her backyard viewing is filled with the brighter objects she can see from home; of those, the ones least sensitive to light pollution and atmospheric haze are double stars.

The list starting on page 34 is one rarely found in modern astronomy books—a year-round double star observing program that can be seen entirely with a 60mm refractor. Thanks to Sissy Haas, the small scope urban observer is no longer left out of the hobby. The multiple stars listed are as easy in a small refractor as the Messier list is for a 6-inch telescope. All the descriptions are based on Sissy's own observations with a 60mm (2.4-inch) refractor.

3
Starhopping

Starhopping is as useful in finding your way around the face of the Moon as it is in locating distant, deep-sky objects. Reference points are different, but the technique is the same: find the bright star (or the big crater) and use it and its neighbors as guides to subtler prizes nearby. If finding an object by its right ascension and declination is the celestial equivalent of looking for a small town by its latitude and longitude, or by the notations in the margins of a road map, then starhopping is like finding the same small town by reference to neighboring cities.

Assessing your sky

Before you begin to search for celestial treasure, you need to know how good or how dreary your sky is. The first, best clue to what is worth looking for and what is hopelessly out of reach is the naked eye's "limiting magnitude."

Fig. 3.1. The skyglow above the suburbs of major cities and above small towns mutes the night sky and hides all but the brightest stars. Jupiter chases Venus down the sky above the Charles River in downtown Boston, Massachusetts. The chart of Orion shows stars to about magnitude 3.5. This part of the sky is shown at the same scale to different "depths" on the following pages.

The limiting magnitude is the brightness of the faintest star you can see on any given night, from any particular site.

How low can you go? Guidebooks assert that the human eye can see to 6th magnitude, an estimate both too conservative and too optimistic. It is too conservative because with practice, a modicum of care, and perhaps a little natural aptitude, most observers can see stars two to three times fainter than 6th magnitude. Exceptional observers under exceptional conditions have demonstrated the ability to see below 8th magnitude. It is too optimistic because skies dark enough hold such a profusion of stars are increasingly hard to find as lighting spreads across the once-dark countryside.

Modern observers in towns and suburbs are limited to about 4th magnitude. Only a few hundred stars shine brightly enough to penetrate the glowing air over well-populated places. If you live under better skies than this, then count your lucky stars (if you can).

Under superb skies, constellations are submerged in veritable drifts of stars. If you first learned your way around suburban skies, then it is entirely possible to be briefly, happily, and completely lost when blessed with really good skies. Excellent skies reveal the most elusive objects and show old friends to best advantage. If making your way through mediocre skies is discouraging, just remember that learning your way around any skies insures that when you do find great skies you will be ready to take advantage of them.

To calibrate their detectors, professional astronomers use small "selected areas." These small standard regions of carefully measured stars typically contain one star bright enough to be visible in ordinary binoculars and a "ladder" of many dimmer stars down to those at the limit of detection with any instrument. The amateurs' "selected areas" are simpler affairs. One is always handy in the northern sky: Face north, find Polaris in the usual way, and note that Polaris is the first star in the handle of the Little Dipper. The Little Dipper is a less perfect dipper than its larger neighbor, and it is fainter. It curls back from Polaris toward the Big Dipper. The stars of the Little Dipper form a useful magnitude sequence ranging from 2nd-magnitude Polaris to a 5th-magnitude star completing the Little Dipper's bowl.

Limiting magnitude is also influenced by how completely dark-adapted your eyes are. It may take an hour or more for your eyes to reach their maximum sensitivity under the darkest skies. Under mediocre skies, it may take only a few minutes. Do not look at any white light while letting your eyes adjust or while maintaining your dark adaptation. To preserve maximum visual sensitivity under good skies, use only a dim red flashlight, and only when absolutely necessary.

Here's a general guide to what you can expect to see under skies with different limiting magnitudes.

Third magnitude: In 7 x 35 or 7 x 50 binoculars, you may spy a few of the brightest Messier objects: the brightest clusters, the Orion Nebula, the innermost regions of the Andromeda Galaxy. Guideposts for starhopping will be few and far between; navigating a small telescope will require perseverance. Deep-sky objects will be undistinguished. This is a sky for planets, the Moon, and double stars.

Fourth magnitude: It may still seem a challenge to find more than a handful of deep-sky objects: bright nebulae threaten to disappear in the sky glow; globular clusters look just like glowing balls without resolution into stars. This level of light pollution yields nicely to nebula filters (like Orion's Ultrablock or Lumicon's UHC). They will help with bright nebulae and make brighter planetary nebulae easy and detailed. Dark adaptation will be casual; five minutes of studiously avoiding bright white lights will give you most of the sensitivity you can use. Binoculars offering a 5mm exit pupil will serve you best here (8 x 40s, 10 x 50s, 14 x 70s).

Fifth magnitude: The Milky Way is noticeable in these skies—in summer it is an eye-catching "cloud" overhead, in winter a subtler band of light running beside Orion. In binoculars and a small telescope, some bright nebulae are detailed, but most are still disappointing. Nebula filters may bring good results. Not only the bright cores but some of the middle regions of galaxies become visible. In small telescopes, globulars begin to get "grainy": their constituent stars may be just at the limit of visibility. Your eyes will reach most of their useful sensitivity in 10 to 15 minutes. Binoculars and eyepieces offering 6mm to 7mm exit pupils begin to come into their own.

Sixth magnitude: The Milky Way is a bright "twist" of starlight in both summer and winter skies—clusters and stars just on the edge of visibility give it a sparkling depth. As twilight deepens, it's easy to mistake the Milky Way for gathering clouds! In binoculars or a wide-field telescope, dark nebulae are recognizable by the absence of stars. Many bright nebulae are easily discernible and show good detail in the telescope. UHC, Ultrablock, and O-III filters aren't needed to see nebulae under these skies, but sometimes they bring out breathtaking detail. Many galaxies in binoculars or a 6-inch telescope are only slightly less extended than in their photographs. You may have to wait half an hour or more for your eyes to take full advantage of the darkness. Binoculars and eyepieces offering a full 7mm exit pupil perform near their best: 7 x 50, 10 x 70, 11 x 80, and other "night glasses" are in their element.

Seventh magnitude: Dark nebulae hang in front of the Milky Way looking exactly like dark clouds adrift in the depths of space. In wide-field instruments,

Fig. 3.2. Orion's stars, down into the 5s.

Table of Magnitudes

-26 The Sun
-13 The Full Moon
-4   Venus
-2   Jupiter
-1.4 Sirius, the brightest star in the sky
0   Vega, the brightest star in the constellation Lyra; overhead in summer evenings in the Northern Hemisphere
1   Arcturus, the brightest star in Bootes (the first bright star beyond the Dipper's handle)
2   Polaris, the North Star
3   Typical urban limiting magnitude (about 160 stars in all the sky)
4   Faintest stars visible to the naked eye from suburbia and small towns in the U.S. and Europe (480 stars)
5.5 The planet Uranus, the four bright moons of Jupiter
6   Traditional limit of visibility on a dark, rural night (+/- 5,100 stars)
7   Faintest stars visible in a very good, dark sky with dark-adapted eyes (+/- 16,000 stars)
8   The planet Neptune
9   Approximate limit of 7 x 35 binoculars
10   Approximate limit of 7 x 50 binoculars (+/- 270,000 stars)
12   Approximate limit of a 3-inch telescope (+/- 2,000,000 stars)
13   The brightest quasar, approximate limit of a 6-inch telescope
14   The planet Pluto
15   Approximate limit of the deepest current computerized all-sky star chart, the Hubble Guide Star Catalog (15,000,000 stars!)
18   Approximate limit of expert observers under excellent skies using the largest telescopes available to amateur astronomers

Fig. 3.3. Rural skies hold stars aplenty: the winter Milky Way appears; M42 is a hazy cloud to the naked eye.

dark nebulae show structure. In summer or winter, the Milky Way itself may be the most striking object of all—one gets the vivid sense that our toe-hold on earth is but a galactic overlook, that most of the natural world is hanging overhead, just out of reach. Familiar star patterns are hard to recognize in the profusion of stars. In binoculars or a 6-inch telescope, galaxies spread out to approximate (or exceed) the dimensions shown in their usual portraits. Bright nebulae show rich detail. O-III filters bring out astonishing contrast in even the faintest nebulae (the North America Nebula glows like a neon sign). Your eyes may take an hour or more to fully adapt under such skies. Do they ever fully adapt, or does your vision improve with every passing hour of freedom from intense light? Even in a 3-inch telescope, deep-sky objects are beautifully detailed; the sight of them in larger telescopes will stay with you for a lifetime. These are the skies we dream of!

The measure of all things

The human hand, at arm's length with fingers spread wide, spans about 20 degrees (tip of the thumb to the tip of the little finger), the basic unit of starhopping. The clenched fist held at arm's length subtends 10 degrees across its four knuckles. The thumb covers 2 degrees (or a little less), and the smallest finger about 1 degree. At arm's length, the smallest fingernail looks a little larger than the Moon.

Test this: your outstretched hand reaches from the head to tail of Cygnus. Cover Deneb with your little finger, and Albireo is under your thumb. It spans the Big Dipper from the end of the handle to the middle of the bowl. Your closed fist at arm's length fits just beside the bright W of Cassiopeia, or comfortably within the Dipper's bowl. Measure the height of Polaris above the northern horizon by stepping off the distance with your hand. The tally should match your latitude.

The field of view of your telescope cannot be so neatly summarized, but it can be easily computed. The "true field of view" of your telescope is a function of its focal length, the focal length of your eyepiece, and the "apparent field of view" of your eyepiece. Eyepieces have very different apparent fields depending on their design. If you hold an eyepiece up (in daylight) and look

Fig 3.4. "Ceiling unlimited, visibility 60 miles. Clearing later," is my favorite weather report from the American West. Here, an endless view in New Mexico gives way to pristine skies at night. Dry air, high altitude, and no artificial lights for many miles promise excellent stargazing.

a particular object it's important to have a good idea of the size of your field of view and an inkling of what you are looking for. Sweeping pays off with novel "discoveries" and can spare you the planning and execution of tedious starhops across difficult passages.

The two techniques can be combined: starhop into the general area, then sweep to find the object you want. Or once an interesting object sweeps into view, back away from it until you find a bright object to mark your place. Sweeping has one advantage over starhopping: the eye-brain combination is especially good at picking up movement. As you move the telescope's field of view, very subtle contrasts are emphasized: objects you cannot see when the telescope is stationary sometimes leap into view when you barely rock it.

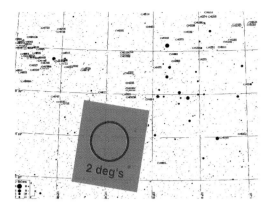

Fig. 3.5. A piece of Mylar or onionskin paper with a circle marking the field of view of a favorite eyepiece or the framing of a favorite lens increases the usefulness of any star atlas manyfold.

What does it really look like?

Photographs whet our appetites for celestial wonders, but they seldom provide us with realistic ideas of what we should expect. Film collects light and piles it up in unfamiliar ways. Photographs of deep-sky objects have more contrast than those objects will actually have in the eyepiece. On the other hand, the eye-brain combination follows turbulent, dancing images and records more detail from lunar and planetary subjects than do all but the very finest photographs.

Digital image processing, long used to bring out subtle details and heighten contrasts, to make "better" versions of traditional photographs, can also be used to make photographs more like visual impressions. On page 53, for example, is the Rosette Nebula, as generally shown in photographs, and then more as you can expect to see it in a superb sky with a wide-field telescope or night glasses. One view approximates the Rosette's appearance without a nebula filter—the star cluster dominates this view and the nebula is only barely visible. Another image suggests what an O-III filter or other aggressive light-pollution-rejection filter reveals. Compare these images to the way the nebula is generally shown in books and magazines in "good" astrophotos made to show the subtlest, faintest details to best advantage. Here, the nebula dominates as it never does visually.

Many objects that appear in photographs with all the apparent solidity of a snow-capped mountain are utterly invisible in the eyepiece. It's not easy to know from photographs which objects are out of reach, which are worth a look on an exceptional night, and which are there for the asking any time the sky is clear. Only experience at the eyepiece will let you judge which are targets of opportunity and which are objects of frustration.

The astronomical menagerie

Galaxies. Except for two or three of the very nearest, galaxies are smaller than their published portraits suggest. They are softer, of lower contrast, and less structured. In the telescope, expect galaxies to glow as softly as the Milky Way does to the naked eye. But remember also that while we never see the center of the Milky Way (because thick clouds of dust get in our way), the brilliant central starclouds of most external galaxies *are* visible. This makes their centers far brighter than their outlying regions and their "surface brightness" much brighter than the Milky Way's. From urban and suburban locations, the central condensations of other galaxies is all you should expect to see. Photographs processed for maximum contrast and sensitivity cause this bright central region to burn out into the more diffuse outer realms of galaxies, making them seem deceptively large.

M31 is close, large, and tilted so its star-crowded center is directly visible. In many ways a typical spiral, M31 is very unusual in the eyepiece. M31's two tiny companions are more typical of the most visible galaxies.

M33 is also huge, but it is traditionally hard to see—actually, I find it easy in even small glasses (9 x 25s, 7 x 35s, 7 x 50s) even under skies with 4th to 5th magnitude naked-eye limits. The key is knowing where to look (see the Cassiopeia key chart) and what to look for (a soft, Moon-sized patch of light, best found by moving the binoculars slightly, faint throughout but softly brighter in the middle).

Incomparably larger are the Magellanic Clouds, satellites to the Milky Way. They are visible only from far southern locations, ideally from the Southern Hemisphere, and look like detached parts of the Milky Way. Each covers a roughly fist-sized area of the sky.

Star clusters. Open clusters are relatively loose assemblies of stars—they form in the arms of galaxies from clouds of dust and gas, and most slowly lose stars to the general galactic population. Many burn with the fierce blue fire of young suns. Globular clusters are more massive, with stars tightly crowded in their cores. A few hundred stars may be individually visible in some nearby globulars; hundreds of thousands more make up their soft background light. While open clusters are among the galaxy's youngest features, globular clusters are its oldest.

At low magnification and to the naked eye, many open clusters look like glowing clouds, like nebulae. In larger instruments the individual stars shine clearly. Some are tiny (NGC 2158 beside brighter, larger, closer M35), while others sprawl over vast tracts of sky. In almost all cases, they consist of the many stars you see and many times more that you don't. Sometimes this unseen spray of fainter stars gives the cluster a deceptive "nebulous" look. And sometimes clusters really are wreathed in glowing dust and gas, leftovers from the clouds from which their stars condensed. Young, blue clusters tend to be "gassy" (the Pleiades, the Rosette) while older clusters are gas- and nebula-free. The Beehive, also known as M44, lies between Leo's outstretched paws. It is so free of gas that large instruments show a few distant galaxies shining right through the cluster.

Unless you know the field of view of your telescope, it's hard to know what to expect when you turn to an open cluster. It's possible to look right through large clusters. The huge (fist-sized) Coma Star Cluster, east of Leo the Lion, is beautiful to the naked eye and in low-power binoculars, but with any more magnification (hence any narrower field of view) it disappears. The Hyades (between Orion and the Pleiades, including the bright orange star Aldebaran) are similarly lackluster in big telescopes but beautiful in smaller telescopes and binoculars.

Under excellent skies, even a 3-inch telescope will show a few globular clusters for what they are: swarms of stars all forming a massive, brilliant ball. Omega Centauri, in the far southern sky, looks as big as the Moon. M13 in the keystone of Hercules, M22 above the Teapot of Sagittarius, M3 in Canes Venatici, and M4 leading the Scorpion's bright red heart are all between ¼ and ⅙ the apparent size of the Moon. Others are only Jupiter-sized in the eyepiece. With dark skies and sharp glass, large globulars can be resolved into 11th-, 12th-, and 13th-magnitude stars. Use plenty of magnification—125x to 200x—for "cluster busting" in small telescopes. In binoculars, in small telescopes, and whenever skies are not dark enough to permit their stars to be seen individually, globulars are soft balls of light, strikingly round clouds of light, sharply brighter toward their centers.

Don't pass up the chance to view a bright globular cluster in a very large telescope at a star party or at public nights at major observatories. In telescopes larger than 16 inches, globular star clusters shine like piles of jewels: rubies, diamonds, and sapphires in extravagant abundance.

Nebulae. Nebulae may be dark or bright, sprawling across many fields of view or so small that only subtle tests can distinguish them from stars. Bright nebulae are distinguished by their origins and the mechanisms by which they shine. "Bright" means only that they are brighter than their surroundings. Some are very dim indeed. Bright nebulae include supernova remnants (the Veil, the Crab), planetary nebulae (the Ring, the Dumbbell, the Saturn Nebula, the Ghost of Jupiter, the Eskimo, the Helix), star breeding grounds (the Orion Nebula, the Lagoon, the Rosette), and reflective curtains (like the Merope Nebula in the Pleiades).

In the telescope, bright nebulae range from practically invisible (Barnard's Loop) to visually striking (the Lagoon Nebula, the Great Orion Nebula). Most respond well to nebula filters of various kinds. LPR (Light Pollution Rejection) filters attach to the eyepiece and reject the scattered light of streetlights and natural airglow to pass only the light by which nebulae shine. Orion's Ultrablock, Lumicon's UHC, and Lumicon's O-III filters all do a superb job. Filters are not a panacea: they exact a certain "photon tax" that may be too

much to pay. Streetlights are effectively turned off, but "continuum
sources" like stars and galaxies are
significantly dimmed, too. Only certain nebulae shine through with
unreduced brilliance. The view is different and sometimes starless in
very small telescopes, which makes
the use of aggressive LPR filters
uncomfortable for some. A 6-inch or
larger telescope benefits most.

Planetary nebulae and supernova
remnants are particularly bright in
the blue-green light of doubly ionized
oxygen (O-III). All the oxygen in creation began in stars that "exhaled" it
into the cosmos through supernova
explosion or shell ejection late in life.
Looking at these nebulae through an

Fig. 3.6. The human hand, fingers outstretched and held at arm's length, subtends
an angle of about 20 degrees from tip of
thumb to tip of littlest finger. Navigators
would use a cross-staff—you have two of
them ready at any time!

O-III filter is like looking through a window on creation—all the elements
heavier than hydrogen (except a little helium) were geysered into space
through such fountains. Some planetary nebulae appear very small—too
small to see as anything more than a starlike point. By holding a nebula filter
over the eyepiece and bringing it in and out of the light path, stars seem to
blink, but the nebula shines steadily.

Your keys to the kingdom

Pattern generating is something we humans do very well: whether in tea
leaves or a scattering of stars, we perceive order where there is none. In a
starry sky, we manufacture lines, circles, triangles, arrows, and trapezoids.
For the most part, these eye-catching patterns do not encompass whole constellations. I am mystified trying to make crabcakes out of Cancer, a goat out
of Capricorn, or fish from Pisces, but Orion does look like a warrior, Leo like
a lion, Cygnus like a swan, Scorpius like a scorpion, and Delphinus like a
leaping dolphin.

On the following pages wide-field photos show how to use the brightest
asterisms of the sky to find some less familiar deep-sky objects. Use my examples without reverence: go and find better routes. Astronomical magazines,
picture books, and especially *Burnham's Celestial Handbook* will suggest innu-
merable treasures worth finding. *Sky Atlas 2000* or the *Magnitude 6 Star Atlas*

Fig. 3.7. The Rosette is a newborn star cluster still wreathed in the gas and dust from which it formed. In photographs, the nebulosity dominates the view. Turn your telescope this way, and you see something else entirely.

Fig. 3.8. At first glance, this is how the field of the Rosette will probably look. All these images are about 3 degrees across, centered on the same point. In fact, they're all the same photograph, digitally processed to look like a classical astrophotograph and to mimic visual impressions.

Fig. 3.9. Under very dark skies, or with an aggressive light pollution filter like an O-III line filter, the nebulosity is apparent but subtle. Use big binoculars or your widest-angle eyepiece.

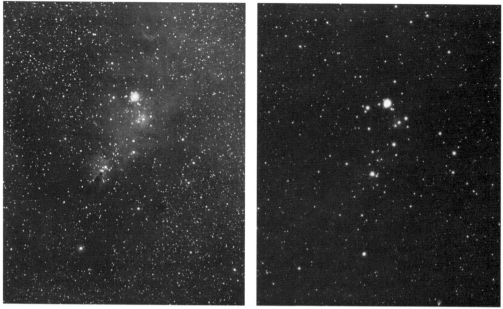

Fig. 3.13   Fig. 3.14

BIG DIPPER STARHOP

"Every pupil should be instructed in the manner of pointing out the North Star at any time of night. If they are enabled to do this, it will assist them in making other important observations, as well as being of use on many occasions which occur in the life of every man. Many persons have been lost in a prairie or other unfrequented places, when if they had been able to have told the points of the compass they could have extricated themselves from their lost situation." Asa Smith, *Illustrated Astronomy*, 1850. Lest you get lost in a modern prairie, practice the following hops:

Fig. 3.15

Fig. 3.16

56

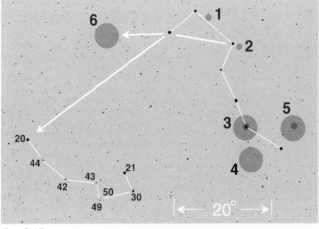

Fig. 3.17

1) M97—the Owl Nebula and its close neighbor "M108," an elongated galaxy. Nebula filters will emphasize the former; only dark skies can help reveal the latter.

2) "M109"—all Messier numbers after 104 are somewhat tainted, having been added to Messier's list by others.

3) Alcor and Mizar—a wide pair in binoculars, and visible to the naked eye; a bright triple in the telescope. See fig. 3.18.

4) M101—the Pinwheel Galaxy. Face on, with a low surface brightness, M101 shines at the end of a short starhop that starts at Alcor and Mizar.

5) M51—the Whirlpool Galaxy. A faint cloud beneath a "tent" of 9th-magnitude field stars; two-lobed in larger binoculars and in small telescopes. In 8-inch or larger telescopes, this is the most vivid spiral galaxy in the sky. See fig. 6.11.

6) M81 and M82—two bright galaxies only about a degree apart. Stars at the corners of the Big Dipper's bowl point to them, just as the end stars point to Polaris. Visible in 7 x 50s even in marginal skies, structured in darker skies. Figure 3.19 shows the galaxies as they appear in large binoculars—except that fainter stars are shown: one of the three closest to the center of the larger galaxy, M81, is a supernova, an exploding star in that galaxy. It vanished in 1993.

The numbers beside the stars of the Little Dipper are visual magnitudes to the nearest tenth with the decimal points omitted. This is one of the amateur astronomer's "standard areas" for assessing sky quality.

LEO STARHOP
Leo is the signature constellation in the realm of the galaxies. Virgo and Coma Berenices each hold far more galaxies but offer no bright star patterns with which to find them!
1) Gamma Leonis, Algieba, an easy-to-find, easy-to-split double star with contrasting colors.

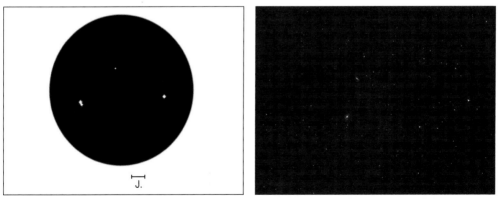

Fig. 3.18   Fig. 3.19

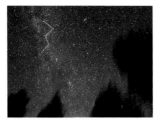

Fig. 3.22

The stars are separated by 4.3 arc seconds and shine at magnitudes 2.6 and 3.8.

2) The Trio in Leo—three bright galaxies below the Lion's haunches. Two, M65 and M66, are relatively easy in binoculars and small telescopes. The third, NGC 3628, is large, edge-on, and dimmed by a huge dust lane. See fig. 6.9.

3) Another small group of galaxies, this one includes "M105" under the belly of the cosmic cat.

4) The Coma Star Cluster. An eye-catching prize in dark skies but invisible without binoculars in marginal ones. Almost any telescope will offer too much magnification—you'll look right through the cluster to empty space beyond!

5) Best found by a short, "micro-starhop" using the southeastern stars of the Coma Star Cluster, NGC 4565 is the finest example of an edge-on galaxy in the sky. Look for a faint needle, about as wide as Jupiter.

6) The Virgo Swarm of galaxies. Bright members carry

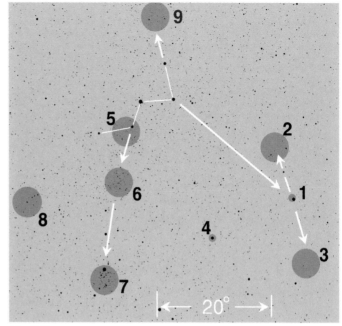

Fig. 3.21

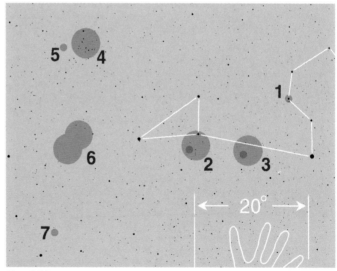

Fig. 3.20

Messier numbers, but to tell which is which you'll need a detailed map with which to go "galaxy hopping." Under dark skies, 7 x 50 or larger binoculars show galaxies everywhere you look in this area.

7) Nothing much to look at, but wonderful to contemplate: 3C-273 is the brightest of all quasars. Even so, it's only 13th magnitude and requires a detailed finder chart. This fount of 1- to 3-billion year old light is a prize for a more advanced starhop of your own devising.

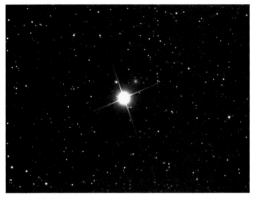

Fig. 3.23

CASSIOPEIA STARHOP

Just as a chance alignment of stars in the bowl of the Big Dipper points the way to the North Star, the bright stars of eye-catching Cassiopeia point to many easy prizes in the northern sky. Use lines drawn by its squashed W to find clusters, double stars, and galaxies.

1) The stars of the W's bigger triangle point to Beta Andromedae. Snuggled up against Beta is a small galaxy and a 9th-magnitude star. The two define a tight, equilateral triangle with brilliant Beta. NGC 404 is visible in large binoculars and small telescopes as a fuzzy "star." Figure 3.23 covers just over one degree.

2) Turn northwest at Beta Andromedae and continue in a line past the next bright star to find M31, the Great Andromeda Galaxy. It's easy even in soupy suburban skies, though you may see only its brightest middle regions. The darker the skies, the more of M31 you can see with the naked eye or with any instrument. Under pristine skies its two companions M32 and M110 are conspicuous. Experienced deep-sky observer Steve Coe takes you on a tour of M31 in Chapter 5.

3) The same distance from Beta Andromedae as M31 but back in the other direction is M33, about which observers differ greatly. I think it's easy, but others cannot seem to see it at all. M33 is a huge, face-on galaxy with a low surface brightness. Its center is reasonably bright and in a star-poor field—look for it in 7 x 50s.

4) Gamma Andromedae, Almach, is a beautiful, easy double star and a devilishly difficult triple. Gamma Andromedae A and B are a brilliant blue and gold pair. The B star is itself a very tight double: B and C are separated by less than an arc second. I've elongated B and C with a 5-inch refractor at 270x: barely and exactly once!

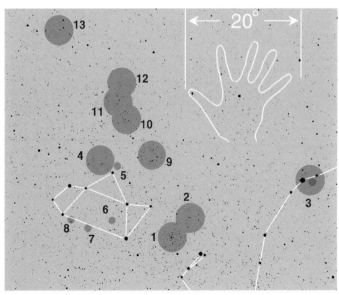

Fig. 3.24

Fig. 3.25

5) M103 is a knot of stars within the W.

6) The Double Cluster—a wild profusion of stars in binoculars, striking colors in a telescope. Just north of the Double Cluster is Stock 2, a similar-sized but much dimmer scattering of faint stars, an echo of the Double Cluster. See fig. 1.6.

7) The Alpha Persei Association—brilliant in any wide-field instrument, including the naked eye. See fig. 1.1.

8) Kemble's Cascade is a chain of 8th- and 9th-magnitude stars that tumbles across almost 4 degrees of sky. See fig. 1.10.

9) M52—just another open cluster in the rich northern Milky Way . . . ho-hum!

SAGITTARIUS STARHOP

The Teapot (fig. 3.25) is an asterism in Sagittarius. You're looking toward the heart of the Milky Way, where stars and deep-sky objects abound and are in dire need of organizing.

1) Think of M7 as the splash beneath the Teapot's spout, a brilliant open cluster to the naked eye and in any instrument.

2) M6, another open star cluster near M7.

3) Antares is the bright red heart of the Scorpion. Antares has a close companion that precedes it across the sky by just a few arc seconds. If you can split Rigel, try tougher Antares. It's possible with a sharp 6-inch telescope and plenty of magnification. Leading Antares across the sky is the large globular star cluster M4, a large, hazy cloud in binoculars. M4 is relatively easy to resolve into stars using a small telescope. North of Antares is the murky Rho Ophiuchi region, which is filled with bright and dark nebulosity. This is visual and photographic fare for only the darkest skies.

4) M22. A tight, round ball in binoculars, a glittering swarm of stars in even small telescopes, located in a rich starfield.

5) M28 begins a parade of lesser globulars down through the Teapot. These are all much fainter than M22. They are suitable binocular targets under dark skies, but tough to resolve in small telescopes.

6) M69

7) M70

8) M54

Fig. 3.27

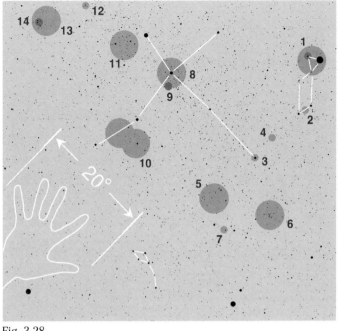

Fig. 3.28

Fig. 3.29

The bright band of the Milky Way angles northward above the Teapot; think of the profusion of bright nebulae and clusters as steam:

9) M8 and M20, the Lagoon and Trifid Nebulae. Under poor skies, the cluster embedded in M8 is all you'll see here. Under better skies, or with nebula filters, the bright gases of M8 are conspicuous. So are dark lanes dividing M20 into three parts, like Gaul, and giving it its name. Star cluster M21 just northwest of M20 is almost overwhelmed. See fig. 3.27.

10) A large, brilliant star cloud.

11) M17—the Omega Nebula, also known as the Swan Nebula. Under dark skies and in a telescope, the bright portions of this nebula look like a swan (or a loon) sitting in a lake of pitch.

12) M16—the Eagle Nebula. Tiny details of this nebula dominate a stunning image made in 1995 by the Hubble Space Telescope showing newly formed stars emerging from columns of dust and gas. The view's not bad with binoculars and small telescopes, either!

13) M11—a brilliant, tightly packed open cluster in Scutum. Easy to any optics in any skies and a naked-eye sight in dark skies.

CYGNUS STARHOP

1) Vega and Epsilon Lyrae. Brilliant Vega is one corner of the "summer triangle" including Altair far down the Milky Way and Deneb, the northernmost bright star of Cygnus. Binoculars split Epsilon Lyrae into two stars; telescopes will split each component into two more (see Chapter 2).

2) Midway between Beta and Delta Lyrae is M57, a Jupiter-sized planetary nebula. Large binoculars will reveal it as a star that will not come to focus; small telescopes show it as a small disk. Larger telescopes, better skies, or nebula filters reveal it as a tiny smoke ring.

3) Extend the line drawn by Beta and Delta Lyrae to find Beta Cygni, Albireo, as beautiful a double star as the heavens offer. Tight in binoculars, easy in small telescopes, with vivid color contrast.

4) Back toward Lyra is M56, a small globular cluster almost lost amid the star-drift.

5) Extend the line from Lyra past Albireo and you come to another planetary nebula. Larger than M57, M27 is a round cloud even in 7 x 50s. Larger binoculars and telescopes reveal interesting structure including "tufts" at either end. It takes its nickname, "the Dumbbell," from the shape of its brightest areas. See fig. 3.30.

6) The Coathanger—Brocchi's Cluster. Fits nicely in the field of larger binoculars and conspicuous even in small ones. Just visible as a bright knot to the naked eye. See fig. 1.11.

7) M71. Another globular cluster, M27's neighbor.

8) Gamma Cygni, Saph. In binoculars, a ring of 7th- and 8th-magnitude stars encircle Gamma Cygni like dancers around a bonfire.

9) M29 is an open cluster set into the ring of dancers.

10) The Veil Nebula. A supernova remnant perhaps 100,000

Fig. 3.30

years old, the Veil is divided into two parts about 4 degrees apart. Each is a faint streak in the eyepiece requiring good skies. Nebula filters work miracles on this object—in larger amateur instruments, they reveal intricate filaments running throughout the Veil. Figure 3.31 consists of three photos: two photos with the 5-inch refractor in hydrogen light show the entire structure; the smaller photo with a 300mm telephoto in white light mimics the Veil's appearance in binoculars.

11) The North America Nebula. Another target for the dark sky, and especially with a filter that isolates the light of doubly ionized oxygen (a Lumicon O-III filter, for instance). In inky skies, this huge and fascinating object is bound north and south by intricate dark nebulae. Under mediocre skies, its outline is doubtful. You need a field of at least 3 to 4 degrees to see it well. Figure 3.32 shows its details; fig. 3.33 is a digitally processed version of the same photo made to resemble my best view with a 5-inch refractor at 22x, with an O-III filter from a 10,000-foot mountain in New Mexico.

12) M39. A bright open star cluster north of the North America Nebula.

13) When viewed in excellent skies, a dark nebula leads away from M39 all the way to the Cocoon Nebula. Can you trace its entire length? See fig. 6.5.

14) The Cocoon Nebula. Under less than ideal skies, the involved cluster is the prize; under black skies, a confusing wad of stars and dark and bright nebulosity nests here.

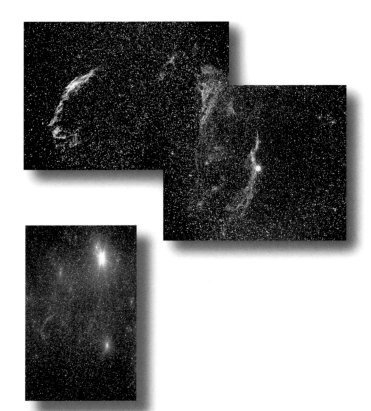

Fig. 3.31

Fig. 3.32

Fig. 3.33

4

Moongazing

The sky is made of stone. . . .
—Jimmy Webb, "The Moon Is a Harsh Mistress"

Two objects in all the night come with lifetime warranties: the planet Saturn and the Moon. The Moon captivates at first sight and rewards every visit. Unlike the feeble, ethereal light of the deep sky, the Moon is clearly something out there, something real, something hands like yours have scooped up, something solidly within the reach of vision and imagination.

Lunar photographs do not promise more than your telescope can show: in almost every case, if some eye-catching lunar detail appears in an Earth-based photograph, it is worthwhile to look for it in person. In surface area, the visible hemisphere of the Moon is similar to North America—with a telescope, perseverance, and practice, it is yours to explore mile by intricate mile.

When the astronauts of Apollo 8 orbited the Moon, they clamped a camera to the window of their spacecraft and let the lunar panorama sweep by while they read a primal story of creation. The unbroken vista of light and shadow, that inexorable parade of mountains and canyons, brings out that kind of feeling. I say that as though I have been there, because a similarly compelling journey awaits the lunar tourist with an eyepiece for a window and a telescope for a spaceship. Sometimes the sky *is* made of stone.

Several times each month, I visit this desert world. In the telescope, without a clock drive, my "point of vision" is about a thousand miles above the dusty lunar plains. The Earth's rotation sweeps the center of my "porthole" across the Moon's severe landscape at the equivalent of about 50,000 mph.

The geometry of moongazing is often dizzying, always instructive. Sir Isaac Newton was both inspired and frustrated by the Moon. The Moon informed his early cogitations on gravity, then defeated his best efforts to account for all its motions.

Consider the waxing evening Moon, whose hard limb curves like an archer's bow aimed at the unseen Sun. Like every lunar "sea," smooth Mare Crisium (the Sea of Crises) yields more subtle details the longer you stare. A few prominent craters on the young Moon hint at the bewildering wilderness of rock yet to come. Wrinkle ridges at the edges of the seas and the bright rays of the crater Proclus suggest the variety of less celebrated features to be found all across the Moon. Domes, rilles, escarpments . . . the Moon is more than a collection of craters!

The long history of telescopic studies of the Moon has left the language of moongazing luxurious. What else "waxes" and "wanes" so many centuries downwind of Chaucer? Where else can we stare at a field of lava and call it, without self-consciousness, "The Sea of Serenity" or "The Bay of Rainbows"? Where but in moongazing are you able, even expected, to borrow such locutions? We owe much of this grace to Father Giovanni Baptista Riccioli, professor of philosophy, theology, and astronomy at Bologna. Father Riccioli endowed the Moon with poetry when he published his lunar map in 1651. Father Riccioli threw out the lunar names of Hevelius, who had thrown out those of van Langren.

Moons are "born" at New Moon, and the position of the sunrise terminator (on the waxing Moon) and the sunset terminator (on the waning Moon) are specified by days past New. This is the so-called "age" of the Moon. The young Moon of evening is the waxing crescent. The week-old Moon is a "half Moon" also known as the First Quarter Moon ("quarter" because a quarter of its entire globe is visible). For the next week, the Moon is waxing gibbous. Full Moon is followed by the waning gibbous phase, which is followed by the Last (or "third") Quarter Moon. The waning crescent rises ahead of the Sun and disappears into the dawn. The Moon is locked in the Earth's tidal embrace: over the eons, its rotation has synchronized with its month-long orbit of the Earth. The Moon always shows us the same hemisphere. The march of day and night account for the changing lunar face.

The First Quarter Moon rises at noon and sets at midnight. The Full Moon rises at sunset and sets at sunrise. The waning gibbous Moon rises before midnight and sets before noon. By Third Quarter, the Moon rises around midnight and stands high in the sky at dawn.

In winter and spring, the evening Moon is well placed above the western horizon shortly after sunset and the Full Moon rides across the sky near the zenith. In summer and fall, the evening Moon hugs the southern horizon. In summer and fall, the waning morning Moon stands tall.

If you pursue lunar observing in depth, this talk of "days past New" and "days past Full" may seem imprecise. The exact aspect of sunlight on a lunar feature can be specified by its "colongitude." More exact angles of lighting on the Moon can be further refined by specifying the angles of "libration" and of "nutation." These two apparent motions are effects of the Moon's steady rotation as it loops through its inclined, elliptical orbit around Earth.

Libration and nutation complicate the moongazer's life: features sometimes appear closer and sometimes farther from the Moon's apparent center. Because of these "motions," sunrise over any particular crater doesn't always come exactly so many days and so many hours after New Moon. The effects of light and shadow are variable from one lunation to the next.

These complexities have an upside, too: libration and nutation allow earthlings to inspect almost 60 percent of the lunar surface. Binoculars can introduce you to the Moon, but a telescope is necessary for really getting acquainted. Virtually any telescope will do: a 2.4-inch, 60mm refractor will reveal more detail than you will exhaust in years; it will even serve better than larger instruments for some specialized projects. Larger, sharper instruments will not be wasted. Knowing the Moon well will take all the time and practice you can spare; the air is never too steady nor optics ever too good for moongazing.

An equatorially mounted telescope with a clock drive or a long-focus Dobsonian on a tracking platform is ideal. If your telescope is undriven, a well-corrected wide-angle eyepiece will make life easier: you'll need to interrupt your observing less often to reset the telescope. With a sharp, wide-angle eyepiece you can use your attention to lock fine detail as it drifts across the field.

A detailed Moon map or atlas is indispensable. Antonín Rükl's *Atlas of the Moon* is hard to beat. You'll also need a detailed guidebook. Michael Kitt's *The Moon* is a systematic, step-by-step guide to lunar features, their appearance, their history, and the science they suggest.

To make a project of learning the Moon, start with broad strokes and work toward finer details. Set your telescope just to the right of the evening Moon, a waxing crescent Moon between three days old and seven. Select a modest magnification, 100x or a little more. As the Earth turns, the brilliant curve of the lunar horizon enters your field of view, curiously drab, almost featureless under the flat light of local noon. White mountains, pale and poetic seas of dust, and the barest outlines of craters follow the shimmering lunar horizon. As the evening Moon drifts through your field of view, noon turns to morning, an inversion of the earthly day. Morning is followed by dawn, by the crisp, saber-toothed shadows of mountains and crater-rims. These stiletto silhouettes make the Moon seem a brittle, cliff-covered world. It is not. Look back at the mountains seen in profile against the sky and in the snapshots brought home by Apollo. The mountains on the terminator are no more rugged than that; sunrise stretches and exaggerates the Moon's soft contours.

Beyond the terminator, dawn gives way to night, but the show is not over. After the hard light of the sunrise terminator comes the pale earthlit expanse of the visible lunar nightside. Many people think it is an optical illusion, but "the New Moon in the old Moon's arms" or the "ashen light" is very real. Finding your way around the earthlit Moon is a test of lunar orienteering. On the earthlit nightside of the evening Moon, look for detail beyond the broad suggestions of "oceans," highlands, and "seas." The bright crater Aristarchus and the dark ellipse of Grimaldi should be easy to spot near the trailing edge of the lunar disk.

On the Moon, Earth's apparent phases are the exact complement of the Moon's phases in our sky. When we see the thin crescent, denizens of the Moon would be treated to a nearly full Earth. At any given phase, earthlight on the Moon is more than 60 times brighter than moonlight on the Earth.[1]

Because the Moon always shows us the same face, the Earth hangs essentially motionless in the lunar sky. The good Earth never rises, never sets except as seen from narrow "libration zones" at the very edge of the visible lunar surface. In the libration zones, the periodic north-south nodding and east-west rocking of the Moon carries the Earth in and out of view. Where earthrise and earthset occur at all, they take days. Our bright blue globe casts soft shadows in the lunar night. You don't have to travel a quarter million miles to experience this lovely light: you can see it any clear night when the young Moon shines in the western sky, or the old Moon rises before the dawn.

Night by night, the sunrise terminator marches westward across the Moon. As Michael Kitt points out, the line of sunrise and sunset moves across the Moon at only about 5 miles per hour! Dawn sometimes leaps ahead of the terminator, from mountain peak to peak, flowing quickly down sloping terrain. But on average, sunrise advances at a walker's pace.

At the edge of night, the subtlest hills and valleys are revealed, sliced as if by a microtome by the hard, slanting light. Three or four days after New Moon, the crescent takes on appreciable detail. For a few brief hours, the backlit crater Petavius with its sharp central peak punctuates the arc of the young Moon. Viewed low in excited evening air, mountain-rimmed Mare Crisium (the Sea of Crises) with the odd, unevenly rayed crater Proclus on its western shore announces another season of detailed exploration.

For the next ten days of every month, learning and relearning the Moon's valleys and canyons, craters and mountains as the Sun rises over them is the usual activity of the Earth-based tourist. Full Moon interrupts this detailed

---

[1]The Earth is 4 times larger in the lunar sky than the Moon is in Earth's, so it offers 16 times the Moon's reflecting area. Every acre of the Earth's bright face reflects 4 times as much light as a similar area of the dusky Moon. Total it up, and the Earth fairly blazes in the lunar sky.

Linger first over examples of that most characteristic landform of the lunar surface: bowls of night, amphitheaters the size of small countries, 50-mile-wide, miles-deep canyons excavated in silence, exploded in seconds from the gray surface by the kinetic energy of worlds in collision, then aged by a billion years of slow collapse under weak gravity and the steady infall of debris from space. Some look fresh, "only" hundreds of millions of years old, others have been filled, their floors flattened, their rims sometimes overtopped by periodic seeps of glowing lava and sudden avalanches of debris expelled from later impacts. Younger craters overlap older craters; lavas lap against the bases of circular walls; smooth plains wrinkle and make domes under the slow pressure of gravity and ancient tectonic forces. It is all a puzzle: in what order, by what processes, has barren stone carried on like that?

Lunar details are small. Most craters are as small in the eyepiece as the largest planets appear—a Jupiter-sized patch of the Moon encompasses the largest of lunar craters. The details that bring the Moon to life are smaller still. The Moon's surface offers far more contrast than the planets ever do, and that helps make small detail easier to see. The comparison should serve to refine your expectations and emphasize the crucial role steady skies, relatively high magnification, and careful inspection play in moongazing.

One second of arc at the distance of the Moon corresponds to 1.2 miles. A 3-inch telescope can resolve features as small as 1.5 seconds of arc on a fine, steady night. With it, you can distinguish lunar features a little under 2 miles across. Larger telescopes, under superb skies, and with practiced observers, reveal features smaller than half a mile. Linear features far smaller than this—hundreds of yards wide but miles long—come and go at the whim of Earth's atmosphere and the nuances of lunar illumination. From how far away can you see a telephone wire with the naked eye? Think how small that is, how it is only the long, drawn-out nature of the object that makes it visible. A lot of the finest lunar details are like that. Still smaller features can be seen when they are struck by the rising or setting Sun. Mountain peaks and bright crater rims glitter against dimmer lowlands. Isolated peaks gleam like stars miles out in the lunar night. Vertical features that wouldn't be out of place on a typical golf course are revealed when they cast long shadows in the slow lunar dawn.

The naked eye can barely distinguish details as small as 1 minute of arc. Objects must subtend 2 or 3 minutes of arc before they are visible as more than just specks. That means that you must enlarge lunar details on the order of 1 arc second by about 120x to 180x. If your telescope cannot resolve detail smaller than 1 second of arc (and that is the limit for a 3- or 4-inch instrument), then any additional magnification is "empty." The view will simply appear dimmer, more turbulent, and fuzzier if you use too much magnification. In a larger, 8-inch telescope, the smallest resolvable detail is about ½ second of arc. Before it can see them, the eye still demands that these details be enlarged to an apparent size of 2 to 3 minutes of arc. 240x to 360x will do so. The atmosphere will not always (or even often) allow you to see detail as small as this, but that's the degree of magnification needed to make your way among the finest lunar details.

The relation between optical limits and the eye's and brain's physiological abilities conspire to make 40x to 50x per inch of aperture all you'll ever need. If you're courting middle age (from either side), very high magnification may serve up another problem: small exit pupils (smaller than a large fraction of a millimeter) will highlight "floaters" within your eye's vitreous humor and other small ocular defects you carry as souvenirs of long life. (Remember that the exit pupil of your telescope is its aperture divided by its magnification.) The bright field of the Moon, unlike the black skies of other stargazing, will showcase these floaters. They can be very distracting.

Crisp nights behind cold fronts are the most transparent, but sparkling nights are not the best for moongazing. "Sparkling" stars mean unsteady air. A close, humid night with "heavy" air and even a very slight haze may bring the best "seeing."

When the light is as you like it on the Moon, it's worth watching patiently even when the Earth's air doesn't want to cooperate. Fleeting moments of steady seeing punctuate long periods of poor seeing. Even if for only five seconds out of every ten minutes, details at the limit of your optics will sometimes snap into view.

Some people are naturals at keeping their bearings on the Moon, but I am not one of them. Here's one way that's independent of the kind of telescope you're using or how many reflections your image has undergone. On the waxing evening Moon, remember that the terminator is the line of sunrise. Shadows point west; the already lit portion of the Moon is east. Conversely, on the waning Moon, the terminator is the sunset terminator. Shadows there point east. Remember north and south by rote: the bright and rugged highlands lie to the south, dark and smooth-floored seas are to the north. If worst comes to worst: squint alongside your telescope, north and south on the Moon are the same as in our sky.

So many Moons . . .

The youngest Moon is just an arc of pale yellow in the evening sky. Like the Cheshire cat, the rest of the rocky world has vanished and only its thin smile remains. The smile of the youngest Moon is incomplete. Until the Moon smooth-floored, dark lava lakes. Alphonsus is the bulls-eye into which Ranger 9 plunged in 1965, sending back photos all the way. Arzachel is the third crater of this trio; it contains a central peak, a well-defined "daughter" crater, and a very fine rille that may resist your most earnest efforts.

In the bright, cratered highlands far to the south, huge Clavius is in full view. Its circular mountain ramparts are rendered elliptical by perspective. Clavius contains several large and many small craters. Now that you know where to find it, make a point to watch sunrise come to Clavius. Keep watch and stay up late about ten days from now to watch the Sun set here.

Also in the lunar highlands, Tycho sits like a spider in the midst of its far-flung rays. Like other young craters, Tycho's inner walls show strong "terraces," and central mountains rise from its floor. Pan north from these subtleties to find the curving ramparts of the Apennine Mountains, the highest mountains of the Moon. At First Quarter and a little after, they lead from the terminator back toward fuller daylight. At their western end lies the sharply defined crater Eratosthenes. (If you see the Apennines curving like a sword, Eratosthenes is its hilt.) Depending upon the exact phase of the Moon, extend the road from the Apennines on past Eratosthenes to huge, yawning Copernicus. (You may have to wait a day for Copernicus—if so, it's worth it!) Its high rimwall, deep center, central mountains, and terraced walls are surrounded by a deep ejecta blanket and countless secondary craters. Focus carefully on the terrain surrounding Copernicus: craters ranging from 4 to 5 miles down to the limit of visibility blanket the

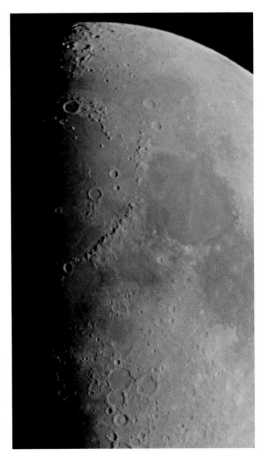

Fig. 4.4. A day after Quarter Moon, features on the northern hemisphere of the eight-day-old Moon announce the arrival of dawn in the realm of lava plains and bone-dry oceans. The greatest range of mountains the Moon offers, the curving Apeninnes dominate the middle of the disk.

area, all of them secondary impacts where debris blasted out from Copernicus rained down.

When Copernicus stands out in early morning sunlight, the waxing gibbous Moon features seas and flooded craters. Toward the northern limb of the Moon is the flooded, flat-floored crater Plato. It looks like a small black eye. If you catch it early enough, Plato's jagged eastern rimwall casts saber-toothed shadows across its floor. A huge block has slumped just inside Plato's western wall. I routinely glimpse a curving line making its way up and around the outside of Plato's rimwall—it looks to me exactly like an access road, curving up to what must be a stupendous overlook. The bright and rugged terrain surrounding Plato and extending both east and west is the lunar Alps. The Alpine Valley cuts a path through these mountains east of Plato.

South of the lunar Alps stretches Mare Imbrium (the Sea of Rains), dark and smooth except for wrinkle ridges and a few isolated mountain ranges. Mare Imbrium's northern "coast" is indented by one of the most dramatic features on the Moon: Sinus Iridum, the Bay of Rainbows. The Bay is an immense, flooded crater; its "shores" are the Jura Mountains rising straight from the lava plains. Promontory Heraclides and Promontory Laplace guard its mouth, and a series of wrinkle ridges look like huge waves rolling in from the Sea of Rains. Sunrise and sunset take most of a day to cross this Bay, but whenever you catch it in low light, it's sure to be dramatic.

Mare Imbrium was the last of the major lunar seas to form, and its creation changed the face of the Moon. The asteroid whose impact formed Mare Imbrium was about 100 miles in diameter. The impact shattered much of the lunar nearside—the Apennines, that massive front, is really just a part of the rimwall of the young Imbrium Basin. The Imbrium Event took place 3,850,000,000 years ago and set off a billion years of increased volcanic activity that flooded the lunar seas and drew the man in the Moon.

As the Moon nears full, the terminator crosses Oceanus Procellarum, the Ocean of Storms. In this far northwestern quadrant, brilliant Aristarchus comes into view a few days before Full Moon. Schroeter's Valley, one of the largest, most prominent of lunar lava-trails, curves around Aristarchus. In the south, Mare Humorum and the crater Gassendi (with its shattered floor) come into view. Finally, dark Grimaldi's appearance at the eastern edge of the Moon announces the arrival of the Full Moon. Use these landmarks as starting points or as destinations. They all reward careful inspection.

Two-headed calves

Tucked here and there are a handful of enigmatic details. If the Moon had national parks, these would be among its lesser ones. (Surely Copernicus would be the Moon's Yellowstone.) These oddities and challenges serve to mark your progress as a moongazer. They are prizes tucked here and there among suspect terrain. They are offered here in no particular order.

Messier A and B, the comet crater. On the young Moon, out in the middle of Mare Fecunditatis (the Sea of Fertility), is our first oddity: the crater pair Messier and Messier B (formerly known as Messier and Pickering and so shown in some guidebooks). The craters are the tip of a bright streak that seems to point them out; the ray and the pair of craters look like a comet pitched against the Moon.

Reiner Gamma. Two days before Full Moon, hours after sunrise on brilliant Aristarchus, the Sun also rises on mysterious Reiner Gamma. Reiner Gamma looks as if it is composed of the same sort of material that constitutes lunar rays. But rather than being strewn far and wide, this material is looped in a small tangle. It does not appear to be associated with any impact structure. If you catch it as close to dawn as you can, you'll see that it shows absolutely no vertical relief. (When the Sun sets on Reiner Gamma, the Moon is a hopeless crescent deep in the morning twilight.) Examining Reiner Gamma in Lunar Orbiter photographs only confirms its lack of vertical relief and lack of association with underlying structures. What has happened here? Proposed explanations involve cometary impacts and bizarre ballistic accidents: could ejecta converging along diametrically opposite trajectories from an impact exactly halfway around the Moon produce an effect like this?

The Straight Wall. On a line midway between Tycho and the most distant point of Ptolemaeus, in the eastern-most lobe of dark Mare Nubium (the Sea of Clouds) lies the Straight Wall. It makes a good introduction to the Moon's finer detail. In the rising Sun at First Quarter, the "wall" casts a shadow; in full sunlight, it disappears. In the low setting Sun of Third Quarter, the west-facing Straight Wall draws a fine, bright line. It's not really a wall. It's actually a long, straight, gentle slope that catches rising and setting sunlight just so. With careful observation, can you

Fig. 4.5. "The moon's a harsh mistress, it's hard to know her well," is never truer than at Full Moon: besides the blinding glare, every familiar feature takes on a new look under the shadowless light of lunar noon.

decide exactly when the "wall" disappears? The Straight Wall sits in the middle of a large circular formation best seen under even lower light. Many very subtle features lie nearby. The prominent crater beside the Straight Wall is Birt. Just on the opposite side of Birt is a very fine rille running parallel to the Straight Wall. When you can discern this rille, you're ready for tougher game!

Stadius. Just before dawn comes to Copernicus, look amid that unseen crater's scattered secondary craters. Back toward Eratosthenes and best seen just at dawn and sunset is the ruined crater Stadius. Stadius seems to be outlined by a spray of Copernicus's secondary craters. If you catch it exactly at sunrise (you have to know where to expect it to appear), the craterlets pocking Stadius gleam just like an open star cluster impossibly embedded in the Moon. This appearance lasts only an hour or so. Under slightly higher light, the accidental sprinkling of craterlets and the wrinkle ridges that subtly define Stadius are visible for what they are. It's hard to believe that the craterlets don't somehow outline Stadius, but frequency studies show that the features are unrelated.

A similarly well-hidden crater lies in the western Sea of Tranquility, where ruined Lamont waits to reward your perseverance. Apollo 11 landed amid the southernmost reaches of the low, farflung fractures associated with Lamont.

Wargentin. Wargentin isn't unique on the Moon, but it is nearly so. On the eastern fringe of the Moon, just before Full phase, "starhop" past Gassendi and Mare Humorum to find huge, rugged-floored Schickard, and beyond it one of the Moon's best jokes. Wargentin started life as an ordinary crater about 53 miles in diameter. Then an asteroid slammed into the farside of the Moon, just around the limb. The impact created Mare Orientale (the Eastern Sea). Only the outer ringwalls of Mare Orientale are visible

Fig. 4.6. Compare the detailed lunar images made under grazing light to the same regions of the Moon under noon light. This illustration may also help you keep your bearings and judge when the march of day and night reaches what point on the face of the Moon.

from Earth as the Montes Cordillera. From space, this mare is a perfect double-ringed bulls-eye. Great gouts of stone jetted across the Moon, cutting down mountains and carving valleys. Several such features can be traced straight back to Mare Orientale. Another souvenir of the formation of the Eastern Sea is what happened to Wargentin. A tide of broken rock and glowing debris swept around the Moon and filled Wargentin to the brim. For no more than a day just before Full Moon (judge when to look for Wargentin by the passage of dawn across Schickard), Wargentin stands out like a crater in reverse. It retains the circular form of a crater, and its rimwall stands out above the surrounding plains. But where you expect to find the familiar, miles-deep chasm, is an expanse of regolith, exactly even with the tops of the rimwall! It's confusing. All the shadows are wrong. A Y-shaped wrinkle ridge crosses Wargentin—it is visible for only a few hours during each month-long lunar day. Pocket Wargentin and its internal ridge as a prize of luck or perseverance.

Linne. Linne is a small crater (only 1.5 miles, 2.4 kilometers in diameter) surrounded by a bright aureole four or five times larger. Linne is about 600 feet deep—a fair match, only slightly larger, for Arizona's Meteor Crater (which remains the only impact structure you can fall into). Linne sits in western Mare Serenitatis (the Sea of Serenity) just beyond the end of the Apennine Mountains. Under a high Sun, Linne is a brilliant white spark; under lower light near first and last quarter, during superb seeing, it is barely visible.

Linne is special as a cautionary tale. Observers as late as the early 20th Century reported dramatic changes in Linne, including earnest reports of its complete disappearance! Photographs from Lunar Orbiter and

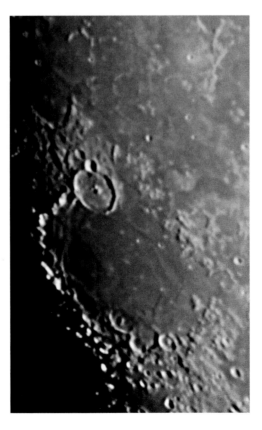

Fig. 4.7. Gassendi is an ancient crater on the shores of Mare Humorum. A complex pattern of rilles decorates its floor. The rilles are not seen in this photo: they are prizes for visual observers with modest telescopes and for lunar photographers with more skill than I have (yet!). Sunrise comes three days after First Quarter; sunset three days after Last Quarter.

the most detailed examination from Earth confirm that Linne is a perfectly ordinary small crater, showing no signs of recent metamorphosis.

The rille in the Alpine Valley. Running down the middle of the Alpine Valley is a hair-fine rille, a severe test for any Earth-based telescope and observer. For a day or two after sunset, look for the rille when the air is still and the Moon seems ready to bear any magnification you try.

A city on the Moon

To hear Baron Franz von Paula Gruithuisen tell it, the Moon is a truly lively place. In 1824, the Baron published a paper describing evidence of lunar habitation. "Discovery of Many Distinct Traces of Lunar Inhabitants, Especially of one of Their Colossal Buildings," detailed what the Baron had seen near the ruined crater Schroter, itself just southeast of Eratosthenes. The region of the Baron's "Colossal Building" is in the hills adjacent to Schroter a short distance back toward Eratosthenes. In fact, the Baron's city *is* the hills! With modern telescopes (and modern beliefs), the Baron's lunar city is next to impossible to recognize.

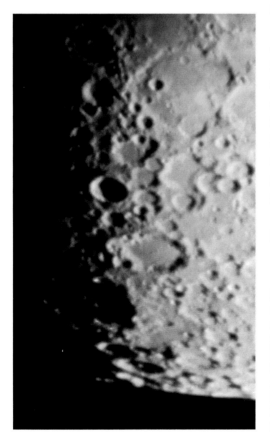

A modern observer, Andrew Johnson, suggests that the Baron's fantastic interpretation is best understood by looking at his "city" four or five hours after local sunrise through a small telescope such as a 2.4-inch (60mm) refractor. Larger, sharper instruments remove too much of the fuzz required for imagination to work its magic.

Before we shake our heads too knowingly about the gullibility that

Fig. 4.8. Tycho's rays have been in view for days before Tycho itself appears. Tycho is the sharp-edged pit near the middle of this photo, just on the terminator. South of Tycho, just barely visible on and beyond the terminator, the floor and eastern rim of huge Clavius is rolling into view. Sunrise is eight days past New Moon, one day past First Quarter. Sunset, predictably enough, is eight days past Full Moon, one day past Last Quarter.

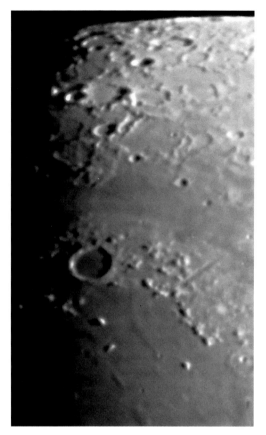

Fig. 4.11. Plato is a smooth-floored, flooded crater. Its eastern rimwall casts stark shadows from wall to wall at sunrise; early risers can see the western rimwall do the same at sunset. East of Plato, the Alpine Valley cuts through the lunar Alps. Sunrise is just hours after First Quarter; the crater is well shown during the eighth day of the month.

south. The "river" kinks along the base of the Apennines. Apollo 15's campsite was beside this rille, between it and the Apennine front, just where the rille strikes out to cross the tiny bay. This is one of the easiest portions of the rille to see and identify with confidence. Autolycus and Archimedes are two large craters in the plains just north of Hadley. Autolycus is the closer of the pair to the Apollo 15 landing site. Between Autolycus and Archimedes, too small to ever hope to see, is a small, metal-lined basin in the Moon's gray dust. No placard signed by presidents, no gold-foil-wrapped launch pad here: this is where the Soviet probe Luna 2 slammed to rest on September 13, 1959. It was the first emissary of Earth to touch another world.

Stupid Moon tricks: occultations

In its month-long orbit, the Moon performs a few special tricks. It slides in front of stars and planets, cutting off their light and marking its own position in space with great precision. Look to the east of the waxing Moon, out beyond its earthlit horizon. If you see a star within about half the Moon's diameter, it may be worth your while to watch as the Moon draws near. If it becomes clear that the Moon will "run over" the star, settle in and keep watch. Don't blink! When the lunar limb reaches the star, the star will disappear with startling speed, with no ceremony and no warning. It's a stark demonstration of the absence of a lunar atmosphere and of just how tiny stars appear. Popular astronomy magazines publish predictions for the occultations of bright stars (and planets) by the Moon; fainter ones are best discovered "accidentally" by checking at the eyepiece.

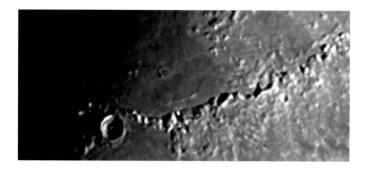

Fig. 4.12. The highest mountains on the Moon, the Apennines curve from Eratosthenes to a small bay where Apollo 15 landed.

By the exact way in which the light of stars dies behind the advancing lunar limb, faint stellar companions too close to see in any telescope are sometimes discovered. A "stepwise" reduction in brightness betrays such faint companions. Such steps may last only a few hundredths of a second.

Before Apollo left its laser-reflecting targets on the Moon, the best way to establish the precise orbit was to time to the nearest fraction of a second when stars disappeared and reappeared behind it. When the Moon passes barely north or south of a star, the star may undergo a "grazing" occultation. Through the eyepiece, it skims along just above lunar lowlands and blinks in and out of view as mountains, crater rims, and even large boulders block its light. I've never seen a grazing occultation but the near miss by one of the bright stars of the Pleiades is a sight I will long remember. With my telescope set to track the Moon, I watched at high magnification as the hard blue star moved inexorably, and with surprising speed, sideways above the dark, yellow-lit mountains of the Moon. Occultation observations are collected and organized by IOTA, the International Occultation Timing Association.

Transient lunar phenomena

A glance at the Moon shows that it been through some violent times. There is no doubt that violent changes on the Moon have been episodic, that there have been both violent and peaceful epochs in the history of the entire solar system. What is not quite settled is how dramatically different these episodes are. Do real changes continue to occur on the Moon?

Apollo seismic stations detected the deep rumbling of a still-living Moon. They recorded the staccato infall of meteorites. One instrument detected the presence of radon gas, which must have been a recent emission, since radon quickly decays into other elements. If you watched, and waited, and took careful note over a period of time, could you see changes occurring on the Moon? The Moon is a continent-sized chunk of terrain, lying open as a map, and—the thought is irrestible—surely in all that emptiness, something *must* occasionally happen.

Perhaps it does. "Transient Lunar Phenomena" (aka "Lunar Transient Phenomena") have a persistent history. Some lunar sites seem given to mysterious clouds, inexplicable brightenings, and rare visual changes. The observed phenomena may be outgassings from volcanic processes, instances of bright fluorescence or piezo-electric glows. They may be dust clouds raised by small landslides or signs of small impacts. Or perhaps they are the celestial equivalent of cabin fever: all that land, so little action.

No one has yet succeeded in unambiguously recording a TLP, but some reports (especially one in 1958 that included a spectroscopic analysis) are fairly convincing. Watching for TLPs is perfectly suited to amateur observers. All you need is time and familiarity with the subject.

The changing light of the lunar day lays a great task at the feet of would-be observers of TLPs. Those who hunt for supernovae check their "crab pots" routinely: they revisit the same nearby galaxies to see if a new star is superimposed on their familiar faces. But what does a TLP look like? Do you note a softening of perceived detail near some well-marked lunar landmark, a lack of expected contrast, a slightly "warmer" hue? These are a few of the forms TLPs take. But they are also the routine tricks of light and shadow offered throughout the lunar day. Separating extraordinary events from the routine play of light and shadow requires exquisite judgment and an expert's practiced eye.

Transient lunar phenomena are often reported (in this context, "often" means every few years) near Aristarchus. Together with its associated highlands, Aristarchus is the easiest small feature of the earthlit Moon to detect, and TLPs are worth looking for in the lunar night as well as in the lunar day.

Eclipses

The newest of New Moons and the fullest of Full Moons are exciting times: one is a total solar eclipse, and the other a total lunar eclipse. The next few years are eclipse-rich, as the following table of total lunar eclipses indicates:

When	Visible from
Sep. 17, 1997	Asia, Europe, Africa, Australia
Jan. 21, 2000	West Africa, Europe, North and South America
Jul. 16, 2000	Australia, east Asia, Antarctica
Jan. 9, 2001	Europe, Asia, Africa
May 16, 2003	Africa, North and South America, Antarctica
Nov. 9, 2003	Africa, Europe, North and South America
May 4, 2004	Africa, Europe, Asia, Australia, Antarctica
Oct. 28, 2004	Africa, Europe, North and South America
Mar. 3, 2007	Africa, Europe, Asia, South America, eastern North America, South America
Aug. 28, 2007	Pacifica, Australia, North America, western South America
Feb. 21, 2008	Africa, Europe, North and South America

Fig. 4.13. Counterclockwise, starting with the large image: the arrow and opening lines represent the direction of view and the limits of the field covered by the high-resolution camera on the Apollo 15 Command Module. The next frame is a digital projection of a portion of the first frame mimicking the view from the command module. The third frame: the Moon up close, sans atmosphere, looking to the side from low lunar orbit. Note the rille curving along the Apeninne front. The rille is visible in 6-inch telescopes when the air is steady and the Sun just up for a day or two (the day after First Quarter is good). At bottom, three snapshots by the crew of Apollo 15 on the ground at Hadley Rille, looking at the Apeninnes from another perspective.

(This table is adapted from Rükl's *Atlas of the Moon.*) Solar eclipses for the next several years are described in Chapter 7.

One of the classic arguments for the sphericity of the Earth is the shape of its shadow on the Moon, and one of the modern values of watching lunar eclipses is the global snapshot they provide of the Earth's atmosphere. During lunar eclipses, the shaded surface of the Moon is lit only by sunlight refracted around the Earth by its atmosphere. The characteristic red light of the totally eclipsed Moon is a composite of all the sunsets and sunrises taking place on the Earth, projected to the Moon and reflected back to us. At such times, when the Earth is a black, orange-rimmed circle in the lunar sky, the Moon is glowing coal in ours.

Following major volcanic eruptions, when the Earth's air is dust-laden, lunar eclipses are generally very dark. The shadow of the Earth is deep and the Moon sometimes seems to completely disappear into it. At other times, the shadow is remarkably light and the Moon remains easily visible throughout eclipse. A more precise measurement of the transparency of the whole

records that today serve as observing road maps for amateurs with tele-scopes Galileo could only dream of.

One of the most unusual logs was kept by French comet hunter Charles Messier—objects that were not comets! In this way he had a list of objects that might otherwise be confused for the comets he sought. Today, thou-sands of amateur astronomers cut their observing teeth on the "Messier list," some in a year-long pursuit, others in one wild spring "all-nighter."

Astro logs may not even be in the form of a conventional journal. Although variable star observer and comet hunter Leslie Peltier left behind a mountain of observational records, his charming book *Starlight Nights* is a legacy of equal importance. Its pages transport us to his Ohio farm in the early days of the 20th century, and his boyhood dream of owning a telescope. The dream is realized only after picking 900 quarts of strawberries to earn the $18 needed to buy a 2-inch scope. Later, Peltier relates his frustration after receiving his first star charts and being unable to find the variable stars he

Fig. 5.1 Ancient cultures of the American Southwest recorded events in their lives by chipping symbols in the rock surfaces. Petro-glyphs like these are com-monplace in the desert southwest and Four Corners area. Some may record astronomical events.

sought. Like most beginners, he didn't know how to relate the field of view in his scope to the chart.

Many who keep astro logs have unfamiliar names, but we hope that through these pages, and through the online community, they will become as familiar to you as the names shining from the past. Who are they? Arizonans Steve Coe and Tom Polakis, Pennsylvanian Sissy Haas, New Yorker Ken Spencer, Ohioan Chuck Gulker, Marylander Gus Johnson, and Virginian George Kelley. And in the future, you!

Why keep a log?

Simply put, a log records what you saw, when you saw it, and how it looked. A year after the fact, will you be able to remember if you saw M6 in the cream soup of the low horizon? If you wrote it down you will. Also, kept over a long enough period of time, a log will serve as a clear record of how much you've grown as an observer. Early notes may detail your frustration in trying to find an elusive galaxy, while later ones show you speeding directly to the target.

A bonus of log-keeping is improved observational ability. If you consistently record your observations you'll begin to see more and more detail. Part of this, of course, is because you're spending more time at the eyepiece. Recording or sketching it, however, forces you slow down. The more time you take, the more you'll see.

Want to challenge your observing abilities? Read the logs of someone else who uses a similar size scope under similar sky conditions. In this world of aperture fever, we often think we need a bigger scope to see certain objects, when the problem may be something as simple as too light a sky, not enough magnification, or not enough time taken at the eyepiece. Reading an experienced observer's notes will give a realistic picture of what to expect from your own equipment.

For example, many people think a 6-inch reflector is too small for deep-sky work. However, expert observer Steve Coe used a 6-inch f/6 Dobsonian and a UHC filter to view the abundance of nebulosity surrounding the Lagoon and Trifid Nebulae (M8 and M20), the dark lanes and streamers from Antares to the Pipe Nebula, and the North American and Pipe Nebulae, which were visible, in their entirety, in the 2.5-degree field of view using a 35mm Panoptic eyepiece. Steve and fellow observer Kevin Gill spend over a half an hour with the North American Nebula alone.

What should you include in a log?

The basics to include in your log are date, time, location, equipment used, and a basic observing note. Although you can note the time as local time, it is a general practice to convert to Universal Time (UT), a standard time-keeping method used in astronomy. To convert to UT, see below. For a more complete log, you might want to add temperature, transparency, seeing, and a sketch. A deluxe astro log might also include other observers notes of the same objects, historical citations, or atlas reference notes. Accomplished lunar observer Chuck Gulker shares a page from his log book.

Targets:	Bullialdus and Kies Pi Dome
Date/Time:	2:00 UT, May 20, 1994
Weather:	49 degrees F, calm, transparency 4/5, seeing 7/10
Equipment:	94mm f7 Brandon refractor, 0% central obstruction, 640mm focal length, Unitron Altazimuth mount
Eyepiece:	Klee 2.8x Barlow, TeleVue Plössl 7.4 for an effective power of 242x or 65 times the lens diameter

Observation of Crater Bullialdus: Excellent view of this terraced walled 61km crater near the terminator tonight. About ¼ of the western inside walls were brightly lit with obvious terracing. Projecting into the northwest was an increasingly wide smooth floor (Valley Bullialdus). Unfortunately, the famous elevated "causeway" was in the shadows of tonight's terminator, thus not observable. Not far from Bullialdus and along the southern portion of the start of the valley, I observed a bright patch. I noticed later in my research that there was a tiny crater Bullialdus L right next to this bright patch which I did not pick up. Now that I know it is there, I bet I could pick it up next time.

Observation of Kies Pi Dome: Observed very easily by "accident" as I zoomed down the terminator. I had to look this up in Rükl's *Atlas of the Moon* (Kalmbach Publishing) to confirm what I was seeing. This is an easily observed circular dome and highly recommended for small scopes. A noticeable shadow projected westward off the dome. No crater pit was observed despite being charted in Rükl. Also noticed a pointy shadow projecting toward the dome being cast by the western wall of the flooded crater Kies.

It's hard to read Chuck's logs without wanting to turn your scope toward the Moon. After all, Bullialdus L and the crater pit on Kies Pi Dome are waiting!

Letter-writing and the astro log

Perhaps in addition to keeping a formal log, you correspond with other amateurs about your observations. If you save your correspondence on the computer, you can always search your files for specific target entries. George Kelley liberally sprinkled one of his letters with references to his attempts to split Rigel, including notes from his observing log. He wrote:

On February 8, I studied Rigel at various powers and found on that particular night I could easily detect Rigel B using only 38x with the AP6. My conviction remains that with steadier skies, Rigel B should be visible using 30x. I prepared

Converting to Universal Time (UT)

Universal Time (UT) is measured beginning at 00h00m00s at Greenwich, England. As you move west from Greenwich you fall behind UT, and must add an hour for each time zone west of Greenwich you live. For example, if you live in San Diego, you are eight time zones away from Greenwich, so must add eight hours to UT. Thus 0400 hours PST is 1200 hours UT. Remember to correct for Daylight Savings.

If you live in:	add
Eastern Standard	5 hours
Eastern Daylight	4 hours
Central Standard	6 hours
Central Daylight	5 hours
Mountain Standard	7 hours
Mountain Daylight	6 hours
Pacific Standard	8 hours
Pacific Daylight	7 hours

aperture reducers for the AP6 providing openings that are 2, 3, 4 and 5 inches in diameter.

The evening of February 24 was clear, but cold and windy. As early as that morning we were still having snowshowers left over from a strong low and cold front that brought tornadoes to Tennessee three evenings earlier. High pressure was rapidly building in.

Stellar images were not good, appearing somewhat bloated. 38x did not show Rigel B. 55x did (40mm Plössl plus 1.8x Barlow). Next, the 3-inch and 2-inch masks were employed. Rigel B remained visible at 55x when either of these marks were used. It is now my conviction that Rigel B probably will be visible if the aperture is reduced more, so to 1.6 inches. Tests with smaller apertures are planned.

The Reverend Webb writes in *Celestial Objects for Common Telescopes,* Vol 2, that Burnham [S. W. Burnham] saw Rigel B with a 1.25-inch refractor and that Franks saw it with a 1.5-inch refractor.

I spent some part of the snowy afternoon of February 15, 1993, preparing aperture reducers for the AP6 with diameters of 1.0, 1.25, 1.5 and 1.75 inches. February 28 brought the first clear evening sky we've had since I constructed the newest aperture reducers. Tallyho and away I go.

Beginning with full 6-inch aperture and the 17mm Plössl for 72 power, I then reduced the aperture to 2, 1.75, 1.5, and 1.25 inches. Rigel B was plainly visible with the 1.75-inch aperture; intermittently visible with the 1.5-inch aperture; and never visible with a 1.25-inch aperture.

When the 1.5-inch aperture is used, Rigel B appears to lie near the diffraction ring

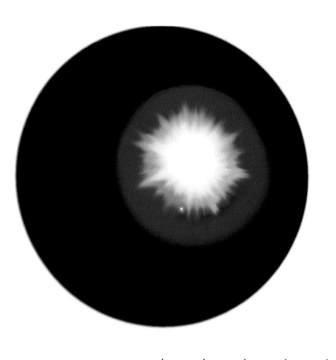

Fig. 5.2. Rigel A and its tiny companion Rigel B. How bright is Rigel? The "faint" secondary star is 150 times brighter than the Sun! The bar indicates scale: tick marks are 10 arc seconds apart and "J" marks the diameter of Jupiter at the same magnification.

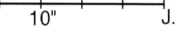

10"   J.

second out from the disk. At times atmospheric turbulence seems to cause the second diffraction ring to devour the image of Rigel B.

Then I determined that 38x is the lowest power I can use and still detect Rigel B. The 40mm Plössl (30x) simply will not separate the images of Rigel A and Rigel B. Rigel B is visible at 38x using both the 2 and 6-inch apertures.

March 1, 1993: With the cirrostrata shield of an approaching storm system appearing low in the western sky, I tried to split Rigel again this evening with different apertures. Rigel B was steadily held using the 1.5-inch aperture, both at 38 and 72 power. The star could be intermittently seen using an aperture of 1.25 inches and 72 power at times of atmospheric calmness.

It is doubtful that an inexperienced observer who is unaware of Rigel B's position relative to Rigel A would notice B's dim transient image provided with the 1.25-inch aperture and 72 power; however it is definitely there, and it is a thrill to share an observation successfully completed and reported by Burnham many years ago.

For George, splitting Rigel wasn't enough—trying to duplicate an observation made (and logged!) by an early astronomer was the challenge. Can you split Rigel with an aperture of 1.25 inches?

Ordinary object, extraordinary log

Arizona observer Steve Coe notes that early on in his log-keeping his notes consisted of entries like "WOW!" or "Beautiful!" He decided to undertake a

Fig. 5.3. M31, the Andromeda Galaxy, as it appears when studied with care under good skies with instruments ranging from 5 to 16 inches. Compare this view with its garish appearance in a typical astrophotograph. Nothing, absolutely nothing, looks in the eyepiece as it does in most photographs! Digital imaging techniques have been used here to adjust the "transfer function" of a single image to mimic both typical astrophotographs and the visual impression.

project of revisiting some of the brighter objects and filling in his notes with more detailed observations. For many observers, the big, bright galaxy M31 (Andromeda) warrants five minutes of awed delight, but for Steve, this visit to the galaxy was different. Why? We'll let Steve tell the story.

In keeping with my plan to re-observe the bright objects and do a good job of it, I spent over an hour with M-31, the Andromeda Galaxy. From a site about 50 miles from Phoenix on a night I rated 7 out of 10 for seeing and transparency, M31 is easily naked eye. It shows a 2 to 1 elongation and is much brighter in the middle. With no optical aid, it is not quite as obvious as the Double Cluster and not quiet as large as the Pleiades.

Moving up to a pair of 10 x 50 binoculars, the galaxy nearly spans the 5-degree field of view and appears somewhat like the shape of a bow tie. It is much brighter in the middle, but seems narrower in the center. M110 is much more obvious than M32. I am amazed at the order in which the companions were discovered.

Using my 11 x 80 finder, the Andromeda Galaxy spans the entire field of view and is much, much brighter in the middle and has a bright nucleus. Both companions are easily seen. The dark lanes are not apparent, but the galaxy cuts off abruptly on the North side. M31 is elongated 5 x 1 in a PA of 30 degrees.

Increasing the aperture to the 13-inch f/5.6 Newtonian shows off this spectacular object at its best. Using a 38mm Giant Erfle eyepiece gives 60x and a 1-degree field of view. The galaxy spans 3½ fields and is very, very bright, very, very large, very elongated, very much brighter in the middle and has a bright nucleus. There are two dark lanes on the North side, but the mediocre contrast of this eyepiece does not do them justice.

Using my 22mm Panoptic provides the best view, in my opinion. The dark lanes are very prominent and that leads to an optical illusion I have occasionally noticed before. It seems that I am looking "through" the dark lanes to see stars on the other side. Of course that is not true and the stars are foreground stars in the Milky Way. The body of the galaxy is smooth, with very little mottling, except around the star cloud NGC 206 at the SW end.

Stepping up to the power of 220x with an 8.8mm. Ultra Wide makes much of the faint outer arm detail disappear, but the core and NGC 206 are fascinating. NGC 206 is pretty faint, very large, elongated 2 x 1 in PA, not brighter in the middle and very mottled. At this higher power this star cloud does not stand out from the galaxy very well.

The core at higher powers is a pretty smooth progression of brighter and brighter inner section of the galaxy, leading to a small nucleus that at 220x and 330x are stellar about 20 percent of the time. There is a 14th or 15th magnitude star about 1 arc min south of the core and this tiny star comes and goes with the seeing.

On previous observing sessions I have picked out the brightest stars and globulars clusters in M31 using an article that appeared in *Deep Sky* magazine.

There is certainly LOTS of fun stuff to observe in the Andromeda Galaxy, it can keep you busy for many nights. I am very glad that my project to re-observe old friends more carefully led me to spend some time with this showpiece.

Following Steve's lead, consider spending more time with the brighter objects; use a variety of eyepieces and filters, then let your notes reflect the detail you see.

Sketching what you see

We non-artists tend to draw what we think we see, not what we actually see. If you're game, spend a few minutes sketching your hand. When you're done, spend another few minutes sketching your hand again, only this time let your eyes tell you what your hand looks like, not what your brain tells you a hand should look like. If your knuckle sticks out, draw the protrusion. If there are four creases across the top third of your index finger, draw them in. Chances are your second picture will look a lot more like your hand than the first.

Drawing astronomical objects takes the same attention. Before picking up a pencil, get comfortable at the eyepiece, and just look. What's the general shape of your target? If it's an open cluster, is it round or angular, regular or irregular? Does it fill up your eyepiece, or just a tiny portion? Is it clumpy, smooth, brighter in the middle? Look. Then look again.

In her book *Drawing on the Right Side of the Brain,* Betty Edwards says the skills we need are not so much art skills as they are perceptual skills. They include how we perceive edges, spaces, relationships, light and shadows, and the object as a whole. Perceiving the object as a whole means having an awareness of what it is, in its totality. If you're looking at a nebula, for example, what does that mean? Is it the kind of nebula that serves as a "star nursery," a birthing ground for new stars, or is it the blown-off shell that represents a supernova remnant? Does it represent birth or death? Is it flowing, looping, intricate? Does it have any relationship to the stars around it?

Paul Rezendes, in *Tracking and the Art of Seeing,* tells us that to excel at tracking animals we need to note the "quality of attention" we give to our habitat, and that the "tracker in the forest is in love with his or her surroundings. In nature, we are open to a larger perspective of self. We learn to walk carefully on this planet. We learn to see it." Like the tracker in the forest, let's become trackers in the sky, and let our quality of attention allow us to perceive. Ready to begin?

Materials:
No. 2 or HB pencil
Artist's stump (found at any art store)
Eraser (preferably with a fine point)
Drawing paper (smooth but not glossy)
Red flashlight (to maintain your night vision)

Drawing a simple star field. Using a compass or another suitable round guide, draw a 4-inch circle. This will represent your field of view. Don't go much smaller than 4 inches, or you'll have a hard time trying to cram what you see into the small space. Begin by sketching the brightest stars. Once captured, they will serve as obvious "anchors" for the dimmer ones. Use the artist's stump for blending hazy edges. It will also help smooth in pencil strokes. Use your eraser to show texture, or to add highlights. Above all: Practice. Practice. Practice.

Drawing the Moon. For anyone who has spent time touring the Moon's terminator, the sheer mass of visual input is overwhelming. Where to even begin sketching? Start with something simple. A simply shaped crater or rille is a good jumping-off place. Avoid craters with complex terraces or tortured mountain ridges.

Begin with making a simple line drawing—nothing fancy, just the basics. Work to include the basic shapes—is the crater round or elongated, are there any breaks in the wall, any small craterlets on the rim, any rilles or rays crossing the floor? Break the sketch down into the most simple shapes possible.

Next, fill in the shadows and erase for highlights. Use the artist's stump to create varying shades of gray in the shadows, then use the eraser to depict sunlit areas. Don't worry about removing too much shade because you can

Fig. 5.4. Even the simplest handwritten logs rekindle memorable observing experiences; for the most gratifying results, mix photographs with your notes. 16mm fisheye, piggybacked, 10 minutes at F4, FujiSuper 800.

go back and add it later. Before attempting to add shadows and highlights at the eyepiece, practice drawing a lunar feature using a photograph.

Go slow, and keep practicing. Don't forget to label your drawings with the time, date, feature, instrument used, and magnification!

For inspiration, Harold Hill's book *A Portfolio of Lunar Drawings* (Cambridge University Press), is unparalleled. His finished drawings, done in india ink, give the feeling of hovering just above the Moon's surface. If the Moon intrigues you, Hill's book is a must for your library.

The electronic log

Using your computer, you can keep as simple or as sophisticated a log as meets your needs. On the simple end, your word-processing program is an

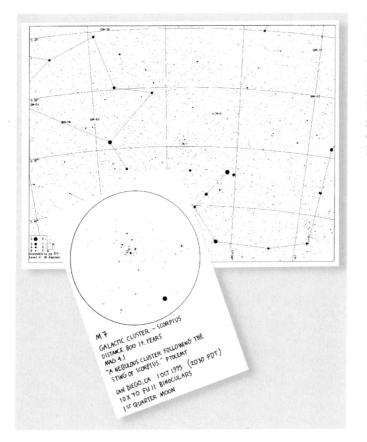

Fig. 5.5. Messier 7 is a brilliant star cluster just east of the Scorpion's stinger. It is the splash below the Teapot's spout. This sketch shows the cluster as seen in large binoculars under marginal skies. The background chart was generated using Bill Gray's Guide program.

easy-to-use logging tool. You may want to set it up by date, instrument, or object. If you'd like to add sketches, print your log pages and put them in a three-ring binder.

If you use a planetarium program, you can keep all your observations using the built-in log feature. Most allow you to click on an object, then go directly to the log option for your entry. Another option is to print a sky map and log of the evening's targets from your software, then add these to your observing notebook.

If you want a more sophisticated astro log, consider keeping your charts, notes and sketches electronically. If you have a scanner (or access to one), your sketches can become a part of your electronic log. And since we all like telling other people about the beautiful things we observe (but have trouble describing them), you can even download photographs from online services or the Internet to accompany your log notes. Both CompuServe and America OnLine have NASA photos available for downloading.

Are you on the Internet? If so, you can look for images using one of the many search engines such as Alta Vista:

Satellite Log

July 2, 1995

I don't believe it. I finally saw *Atlantis/Mir* at 8:23 p.m. (PDT). Every attempt at a dark sky and a well-placed pass have been clouded out. Since the Sun set only 20 minutes earlier, and the sky was still very bright, I doubted I could pick out the pair. Using the Traksat star plot they would pass Altair, but it was so light I couldn't see Altair. I looked where I knew it should be (very, very low on the Eastern horizon), and there they were, the shuttle and the space station, racing from the northwest to the east. I read the online estimates of magnitudes of -5 to -6, which must be true, based on my being able to see them even in the light sky. What a thrill to witness history!

http://altavista.digital.com/

or one of the World Wide Web addresses below. Please note that the Web is evolving at such a rapid pace, all of these addresses may not be correct by the time of publication.

Jet Propulsion Laboratory Home Page

http://www.jpl.nasa.gov

NASA Home Page

http://www.gsfc.nasa.gov

Comet Shoemaker-Levy

http://seds.lpl.arizona.edu/sl9/sl9.html

National Optical Astronomy Observatories

http://www.noao.edu/noao.html

Johnson Space Center

http://images.jsc.nasa.gov

These sites are provided as a jumping-off place to other astronomy-related pages, and are by no means a complete list.

Taken a step further, you can even share your own astrophotos with the global community. Deep-sky observer Tom Polakis has created his Home Page on the World Wide Web. Tom's page contains beautiful images, not only of deep-sky objects, but of the natural world and its place in the astrophoto. For more of Tom's photos and sketches, see his excellent article on observing the Herschel 400, in the June 1995 issue of *Astronomy*. To access Tom's Home Page: http://www.indirect.com/www/polakis/skyphoto.html

The last word

Astronomy logs are an invitation to friends, both near and far, to come into your cosmic living room for a night of shared vision. Although none of us saw M31 through Steve Coe's eyepiece, each of us who read his log shared his experience, and in the doing were there. There's something personal about keeping a journal of all we see, because it is so much a part of who we are. In sharing our skies, we share ourselves. Thanks to Ken Spencer, you are now invited to share his journey to the stars.

For the first time in a long, long time we are having clear weather during the dark period of the month. I called some people from the club yesterday but because it was Sunday no one wanted to drive the hour and a half out to the East end of Long Island to our observing site, where you will find some of the darkest skies on Long Island (which isn't all that dark). I have discovered that I really need to spend time under dark skies. I love seeing the Milky Way arc from horizon to horizon, and watching the stars sparkle like diamonds on black velvet. It's just not the same if you are looking up at a gray sky and the Milky Way is just barely visible, if at all. I have found I need to be out in the open, in the dark, and just experience the night and the beauty of the heavens.

So off I went, alone, to the East end. I brought my 6-inch f/4 RFT, which gives spectacular wide field views of the sky. After I built that telescope I stopped using my 10-inch f/4.8 for almost six months because the views in the RFT were so spectacular! It was a completely new experience for me to scan up and down the Milky Way in Sagittarius and be astounded by the beauty of the fields full of stars, in clumps and wisps, strung out, it seemed, endlessly.

When I got to our site I was the only one there. I set up both my 10-inch and the 6-inch and set to work. I spent almost 30 minutes looking for Uranus but with no luck. I tried last month scanning around an area from memory, but without the finder chart there was little chance I would find it. It looks just like another star at low power. If you find it and use high magnification it becomes a little green disk. The act of actually locating Uranus and Neptune is actually more exciting than seeing them once they're found. So this month I brought the Observer's Guide (Royal Astronomical Society of Canada), with its highly detailed finder chart. However, since it covers a VERY small area of the sky (about 1 x ½ degree) I was not able to orient the chart to the larger part of the sky I needed to look for first, the brighter stars, then the fainter ones used to find the planets. Some nights you can't win! Next month I'll bring ALL the star charts!

There I was all alone, no distractions. I love to observe with other people, but sometimes it can be too distracting. With no distractions I can concentrate on a more organized observing session, relearning some of the faint, and less familiar constellations and some unfamiliar or forgotten Messier Objects. I spent several hours doing just that. Even though it can be frustrating locating a difficult new object, it's always rewarding to finally have it slide into the field of view.

I have a folding beach chair that I use with my 14 x 70 binoculars and a binocular arm. But some nights I may spend half an hour or more just lying back in the chair, without any optical aid at all, looking up at the heavens. That part of the observing is as rich and rewarding as hunting down the "faint fuzzies" with the 10-inch, which is a big part of my usual observing routine.

It was really a rewarding evening and I saw a bunch of new things. And of course there are the old standbys you always seem to look at first thing, and then again at the end of the evening's session because they are so beautiful and because, somehow, they seem to "anchor" me to the sky—old friends who never change: The Double Cluster in Perseus, the Andromeda Galaxy, the Veil Nebula in Cygnus, the Pleiades, and of course the planets Saturn and Jupiter. Being out there was restoring for the soul. A rich and wonderful night. After a night like that, all's right with the world!

6
Photography Under The Open Sky

Johnny Horne is a professional photographer and amateur astronomer whose night work is among the very best. One of his presentations to amateur gatherings is a humble show chronicling his years-long struggle to make the night sky a natural subject for his cameras. His collection of mistakes ranges from the minuscule to the ludicrous. It is unique only because it is preserved and presented with high-spirited humility. Everyone with even a single astrophotograph on the wall or in a magazine has a hundred (or a thousand) others that are embarrassments. Keeping them around, even if safely out of public view, is important. Photos that are discouraging when they emerge from the developer soon become reassuring: only in comparison to earlier efforts is slow progress made plain.

Every snapshot of a colorful sunset can be regarded as an astrophotograph. So can a four-hour exposure with a solid-state detector attached to a custom optical system. In between is a full spectrum of photographic opportunities. Think of astrophotography as nature photography by "barely available" light. A few simple tools make astrophotography possible, and a few more make it casual. Beyond that lies blissful obsession.

No matter how ambitious your photographic goals, a little experimentation with a camera and tripod provides important information. A stationary camera reveals how bright the sky is above your observing sites and helps define how sharp is sharp enough. It hones intuitions and expectations; it "calibrates" your ambitions.

Fixed camera work introduces the puzzle of how film behaves under the stars. Color films come and go, and their behavior in feeble starlight changes in unpredictable ways. Most (perhaps all) off-the-shelf film "slows down" with long exposures. A star photo exposed for 10 minutes will not be twice as dense as one exposed for 5. This "reciprocity failure" sets in when exposures extend beyond just a few seconds. Most of the time when our aim is casual images of the night, we muddle through with fast films working at a fraction of their rated speeds. "Hypersensitizing" minimizes reciprocity failure by specially preparing the emulsion for long exposures, but hypered film is expensive, fragile, and poorly suited to ordinary photography. There's no reason to use such special film for your first efforts and no need to put off getting started because the film in your camera isn't tailored for starlight.

While very fast film (such as ISO 1600 or 3200 color negative films) is needed for some star photos made with stationary cameras, less radical color films (ISO 800, 400, 200, or even slower) work well for gatherings of planets, auroral displays, and noctilucent clouds. Fuji's ISO 400 and 800 color negative films are excellent choices for virtually every astrophotographic purpose. They are sharp, forgiving, and saturated. They respond well to exposures of up to 20 minutes without hypering.

Kodak's faster varieties of T-Max and many Ilford films work well for casual black and white astrophotography. For deep-sky photography through piggyback cameras and at telescopic prime focus, and especially for escaping city light through the hydrogen-alpha window, no film compares to hypersensitized Kodak Technical Pan (without hypering, TechPan is excellent for the Moon and planets and for earthly landscapes). When you feel the need, hypersensitized films are available from Jack Marling's specialty supply house, Lumicon. With experience, you may decide to invest in the paraphernalia to hyper your own for the sake of economy, experimentation, consistency, freshness, and control.

If you decide to start with color print film, for the convenience of off-the-shelf availability and street-corner processing, be sure you can work with your local one-hour lab or custom photo shop to get the best results from your efforts. Some films record the sky as pale magenta, others as green. Very few can be relied on to produce a neutral gray or a rich, black "background" (the glow is really a foreground, of course, but it's hard to think that way while looking at any astrophoto). It's easy to correct these distracting color casts in the printing process if you make your own prints. Having a helpful technician behind the counter can be almost as good: most automated printing equipment can be easily adjusted to change color balance and density, so visit labs until you find someone willing to help with your unusual images.

Copying negatives onto Kodak 5072 reversal film (a C-41 film sometimes called Versacolor-III) turns them into rich transparencies. Filters used in color printing correct (or produce!) many odd color balances while turning negatives into slides. Judicious use of the same filters when shooting near duplicates can turn peculiar slides into better ones. Remember that your choice of a film is not forced by whether you want slides or prints (or digital files) as your final product.

Copying slides increases their contrast and brings out faint objects invisible in the originals. Do it yourself, or have a lab do it. Duplicates are also more relaxing to handle: you won't lose a unique two-hour exposure purchased with a week of annual leave when duplicate slides get scratched.

Kodak PhotoCD conversions offer another means of tweaking images. Digitizing images and loading them into a desktop computer provides a level of control previously undreamt of. Digital images can be improved in ways even professionals could only wish for a decade ago. And, of course, they can be bastardized in new and wonderful ways, too. Outright fakery has never been easier: the border between what is real and what is "enhanced" deserves careful consideration. The driving aesthetic in astrophotography, the ideal by which all images are judged, remains the super-contrasty images made for science in the mid-20th century on Kodak 103 films and plates. Such images were not made to be pretty or to accurately depict what the eye had seen. Just the opposite! Photography revealed things the eye could never see. Perhaps digital techniques will eventually permit (and popularize) astrophotographs that look much more like what is actually seen in the eyepiece and that render unseen wonders much more gently.

Stage 1: Astrophotography with tripod and camera

A tripod, a 35mm camera that can be set for long exposures, and some sensitive film are all you need to get started. A normal lens will do, a wide-angle lens will do better, and a fast wide-angle lens will do best of all.

Everyone knows the Earth spins on its axis. When using a camera and a tripod, you'll have to decide whether the turning of the Earth is a problem to overcome or a dramatic opportunity to seize. It's a problem if you want to use fine-grained ISO 200 to 400 films to make photographs of stars looking like stars. It's a boon if you want to turn stars into colorful, dynamic streaks. Make star-trail photographs using long exposures with a fixed camera. A typical star-trail exposure on a moonless night is an hour or two at F8 on ISO 200 to 400 film (the small recommended aperture helps combat sky glow). Earth's rotation carries stars through an arc of 15 degrees for every hour the shutter is open. Star trails can be as long you like provided only that the sky is dark and the night lasts long enough.

Even so, capturing star trails is not always easy. Airplanes wander through your frame with running lights gleaming and strobes flashing. Dew on your lens makes star trails gradually thin and disappear. Clouds ruin the

45-second exposure (1000/24 = 42) before stars blur into noticeable trails. This is the way to use a stationary camera to photograph bright Orion and many fainter fields as well.

Leaving the shutter open longer than about twice the time allowed by the rule of 1000 will not always reveal fainter stars. During long untracked exposures, light from stars drifts across fresh, unexposed parts of the film. It doesn't accumulate photon-by-photon on one particular spot.

The Dippers in the north, the Cross in the south, and other objects at similar declinations make the easiest targets for your stationary camera. Exposures can be up to twice as long when looking this far away from the fast-moving celestial equator (where Orion and the center of the Milky Way shine brightest). Some scenery within easy reach of the stationary camera: bright comets, auroral displays, gatherings of planets, ice-blue noctilucent clouds, satellites and meteors, the zodiacal light, the shadow of the Earth rising from the east just after sunset or settling in the west just before sunrise.

If you have a sensitive light meter (or an automatic camera) and you are aiming into bright twilight to capture a planetary grouping, measure the light in the brightest part of your frame. If the meter suggests less than a couple of seconds, the best guess is likely to be half or a fourth of the indicated exposure. If the suggested exposure is several seconds, believe the meter and let reciprocity failure insure a darker sky. Such intentional underexposures will deepen the twilight's blue and leave bright planets shining brightly. Just to be safe, "bracket" widely, up to two or four times the measured exposure. Your eyes and your brain compensate as the light fades, and even after years of practice you may be a very poor light meter in rapidly dimming or brightening light—even the best guesses can be very wrong!

If your aim is the deep sky, then the spinning Earth presents a serious obstacle. The 30-second to one-minute limit it imposes even on exposures made through wide-angle lenses means that to photograph rich starfields with a stationary camera, you must use very fast film (ISO 1000 or faster) and reasonably fast lenses (F2 or faster). Fast film and fast lenses both come with compromises: such film is grainy, sometimes hard to find on short notice, and is more vulnerable to X-ray machines, age, and heat than ordinary film. Fast lenses, as a rule, are not nearly as sharp when used at full aperture as when closed down a stop or two. It may seem absurd to pay dearly for a 35mm F1.4 lens only to use it at F2.8, but remember that the less expensive 35mm F2.8 lens would also have to be stopped down to be sharp. You still benefit from faster lenses even if you can't always use them wide open.

Roger Ressmayer is a name to watch for among practitioners of the art of astrophotography using a stationary camera. *Astronomy* and *Sky and*

Telescope routinely carry examples of what can be done with practice, diligence, and skill. To record much fainter sights than those you can see with the naked eye, or to capture the starry night on finer-grained film, and still stop the stars in their flight, you need another accessory.

Stage 2: Astrophotography with a barn-door mount

Stopping the sky requires an equatorial mount. This can mean either a substantial investment in precision machine parts or something as simple as a barn-door, or "Scotch," mount. The former may cost as much as a good used car (and seems to weigh even more during setup). The latter is yours for spare change and weighs just several ounces.

Both pieces of equipment have their places, but only the barn door's place is in your camera bag, tucked between two lenses and your strobe, ready for casual starshooting. Extraordinary astrophotos can be made using a compact barn-door mount built from parts found in any hardware store.

The mathematical principles that govern barn-door mounts are important only if you mean to viscerally understand their limits and design mounts that meet your particular needs. The underlying math also reveals why some of us who routinely use barn-door mounts also park a full-sized equatorial in the garage. Their simplicity and slightly irreverent name belie a venerable history. The barn-door mount is called that for at least two reasons. One is that its motion is as simple as that of a barn door swinging on its hinges. The second has to do with Dr. Donald Menzel's penchant for chasing total solar eclipses to remote corners of the world. Throughout the middle years of the 20th century (when air freight was neither as convenient nor as cheap as it is now), Harvard's Donald Menzel transported forests of telescopes and spectrographs into deserts and jungle clearings, where he commanded them to track the Sun. It was hardly practical to carry and operate separate equatorial mounts for each instrument, so he placed them together on large tilting platforms—platforms about the size and shape of barn doors. To stop the Sun for all the instruments on one of his "barn-door mounts," Menzel used a screw jack to tilt it slowly through a small arc. Barn-door mounts work every bit as well as full-featured equatorial mounts—but only for a few minutes at a time. Because no total solar eclipse lasts more than 7½ minutes, such simple mounts were perfectly adequate.

In 1985 and 1986, Halley's Comet (the tenth or twelfth "event of the century") captured observers' imaginations. The need for an easy-to-carry, inexpensive "equatorial for the masses" popularized the barn-door mount. By the time Halley came by, films had improved so much that deep-sky photographs

Fig. 6.2. The Milky Way above the Kaibab National Forest south of the Grand Canyon; 12 minutes at F3.5 with a 16mm fisheye lens on 1600 ASA Konica print film. The barn-door mount was advanced every 20 seconds. M8 and the star clouds of Sagittarius were made with the same mount using a small electric motor to turn the screw continuously: 135mm lens, 20 minutes at F2.8 on Konica 1600.

could be made in just a few minutes. The barn door enjoyed a renaissance as comet-watchers scrambled southward to see Halley in dark skies, sometimes half a world from home.

To see how these mounts work, hold a book at eye level. This one will do fine. Tilt the book so that its spine points upward at about 45 degrees. Sight along the book's spine and imagine that it is aimed at Polaris. Open the book so that its front cover moves through an angle of about 15 degrees. It's easy to see that if a camera bolted to the front cover were aimed due east and the book were opened at exactly the right speed, the camera would remain aimed at a rising star. With the camera's shutter open for several minutes, you'd get no star trails, just the slow and steady accumulation of starlight on film. It doesn't matter where in the sky you aim the camera: as long as it shares the motion of the opening book, the motion of a barn-door mount, the camera remains in step with the spinning sky.

Since Halley, several elaborations of this simple design have been sketched and built. Multiple-arm designs by David Trott answer virtually every defect inherent in the mount described here. Other compact tracking devices—the d'Autumn platform and the Poncet table—came to prominence

Fig. 6.3. Detailed view of a camera on a barn-door mount; looking south.

about the same time. To take advantage of any of these ultra-precise designs requires much more care in construction and the added paraphernalia of a motor and power supply. Such elaborations are left as projects for interested readers; I prefer to keep things much simpler.

Using a barn-door mount

In practice, the axis of the hinge needs to be aimed only as closely as is convenient at the north celestial pole. Exposures will be relatively short, and many other small errors will be present, including the long "latent" periods between turns of the screw. Taking squinty-eyed aim on Polaris, itself almost a degree from the pole, is close enough.

The rule of 1000 described above tells you how long you can wait between adjustments of the barn-door mount without producing noticeable star trails. For exposures of up to 20 or 25 minutes, the "latent" period in the barn door's operation is the only significant source of stardrift. A 50mm lens can be used on mounts to which you give a twist every 15 or 20 seconds; a 35mm or wider lens can be carried by a mount turned only every 30 seconds.

Load medium or fast film in your camera (ISO 400 to 1600). Put it on the ball-and-socket adapter and choose your first subject.

In the northern hemisphere, start with the bolt fully retracted and the hinge nearly closed. Open the shutter, wait 15, 20, or 30 seconds (depending on the lens you've selected and the design of the mount), then give the bolt its periodic half- or quarter-turn to bring the stars home. Keep it up till the exposure is finished.

Each quarter turn of a 32 pitch bolt advances the bolt just 0.0073 inches; it opens the hinge by 3 minutes and 45 seconds of arc. You have borrowed quite a lot of precision from common hardware! On average, you can judge a quarter turn very closely. Note also that any errors you make are likely to average out rather than accumulate over the duration of the exposure.

A small motor can be cobbled onto this mount to turn the bolt at a slow, steady rate. A motorized barn door can carry a short telephoto (up to about 135mm) with very good results. But if you want to go to the trouble to motorize your mount, consider more accurate, more complex designs that do not suffer from the inherent limits of this one.

Stage 3: Piggyback astrophotography

Long, deep exposures using telephoto lenses require an equatorial mount with a clock drive and a guide telescope to insure that the camera's aim remains bolted to the stars. Cameras can be "piggybacked" on virtually any telescope. Every major equipment manufacturer (including Celestron, Meade, and TeleVue) offers a means of mounting a camera alongside their telescopes. If a piggyback adapter for your telescope is not available commercially, or if you'd just prefer to make your own, the only important considerations are obvious ones: the camera must be held rigidly so that there is no perceptible play between the camera and the telescope; it must not interfere with the motion of the telescope; and it must not make the telescope impossible to balance. It is best if the camera can be maneuvered independently of the telescope to permit alternate compositions and to allow the telescope to hunt for a bright, nearby guide star, but this is not strictly necessary. (It can even be a problem if carried to an extreme: aiming the camera far from the direction in which the guiding telescope points leads to "field rotation" whenever guiding corrections are applied, and stars in your pictures become star trails centered around an unfamiliar pole.)

Some ball-and-socket camera mounts are robust enough to hold a camera and short telephoto lens rigidly through a long exposure, but most are not. The camera can be mounted on any part of the telescope tube or on any piece of the equatorial mount that shares the telescope's motion. For this purpose, fiberglass telescope tubes lack rigidity. Reinforce them carefully before attaching a piggyback camera or (better) find a means of attaching the camera directly to the metal mount.

Piggyback photography with wide-angle and normal-focal-length lenses is the closest thing astrophotography offers to snapshooting. Align the mount within at most half a Moon's diameter of the pole, focus your lens at infinity, set the aperture a stop or two below wide open (unless you have experimentally verified that the lens works well at maximum aperture), and open the shutter for as long as the sky will allow. Voilà!

Piggyback photography using longer lenses becomes proportionately more difficult, and using telephotos with focal lengths of 300mm and longer

Continued on page 113

Barn-Door Principles and Design

The simplest barn-door mount is built around a mathematical approximation: at small angles (up to about 1/10 radian, about 6 degrees), the angle in radians is equal almost to its tangent, and the tangent almost to the sine. For these small angles—which means for exposures only up to 20 or 30 minutes—the effects of this geometric approximation can be ignored. Advancing a bolt through the bottom plate at a constant rate moves the upper plate at an almost constant angular rate—in fact, as the angle increases, the angular motion of the mount decreases. The mount is always falling slightly behind the sky, and falling behind it faster and faster as time goes by.

With respect to the stars, the Earth spins at a steady one rotation per sidereal day, once around the clock in 23h 56m 04s. That's 360 degrees in 1436.07 minutes, or 0.2507 degrees per minute. Reducing this to radians, the Earth turns 0.00437 radians every minute. So the barn-door mount must open by 0.00437 times the distance from its axis of rotation to the drive bolt in the first minute. We want to position the drive bolt so that one turn produces this movement. Where does the bolt go? Consider a 1 RPM mount made around a 32-pitch bolt: 1/32 inch per turn divided by 0.00437 equals 7.14 inches, as shown in the table below. In the next minute, the Earth will turn the same 0.00437 radians; another twist of the mount's drive bolt will advance the drive bolt the same linear distance, but the mount's jaws open through very slightly less angular distance. In the second minute of your exposure, the inherent error is only one part in 20,000. But that error accumulates from minute to minute at an ever-increasing rate.

This built-in error limits this simple design to short exposures. If you know that you want to make very long exposures—up to an hour—with this barn-door design, you can build the mount with a slightly shorter base than specified in the recipe below. If you make the base 2.5 to 3 percent shorter, your mount will run slightly fast at first. The drive will first overtake the sky and then begin to lag. Stars will trace a small arc while the mount runs fast, and then retrace the same small arc while the sky overtakes the slowing mount. Naturally, such a short-armed barn door will work best with wide-angle lenses. In an hour, its accumulated error will be similar to the trailing captured by a stationary camera in 30 seconds (or even a little less if you manage to build it to perfection). A 35mm or shorter lens could make a perfectly acceptable one-hour photograph on such a mount.

As with star trails and perceived sharpness, your personal judgment (and your patience) will determine the outer limits of the barn-door mount. More complex designs with cams or multiple lever arms can reduce design errors to truly insignificant levels.

The hemispheric cap nut on the mount I use is in effect a simple cam. When using a cap nut, reduce the calculated base dimension by about 0.5 percent for optimum results. If you can't place the bolt that closely, just do the best you can: aim for the nominal measurement but be sure to err on the short side. Building barn-door mounts is a casual, but not a slovenly, affair. Errors on the order of ¹⁄₁₆ inch are harmful more in theory than in practice.

The inherent error in the simple design shown here, modest errors in alignment, small errors in construction, and the staccato, imperfect motion produced by manually turning the drive bolt a few times each minute combine to produce tracking errors comparable to those we judged acceptable from exposures lasting a few to several seconds with a stationary camera.

Star trails resulting from stationary camera exposures provide a convenient yardstick by which to compare barn-door tracking performance. In the tables below, "trail length" is the inherent design error expressed in terms of the length of an exposure (in seconds) with a stationary camera that would produce an equivalent amount of trailing.

The first table does not consider the effect of the cap nut. Instead, assume that the drive bolt is filed to a blunt point that pushes up under the top of the hinge:

		Trail length after							
Drive Bolt	**Base**	**5m**	**10m**	**15m**	**20m**	**25m**	**30m**	**35m**	**40m**
¼ x 20	11.428"	0.0	0.4	1.2	3.0	5.8	10	16	24s
³⁄₁₆ x 32	7.142"	0.0	0.4	1.2	3.0	5.8	10	16	24s
1mm	228.6mm	0.0	0.4	1.2	3.0	5.8	10	16	24s

Using an approximately spherical cap nut with a radius of 0.2 inch, the barn-door mount performs slightly better. Here's why: As the jaws open, the point of tangency between the cap nut and the upper plate moves "down" the cap nut, effectively shortening the base and speeding up the angular rate at which the mount opens. This partially and temporarily offsets the inherent error in the design.

In the next table, the base dimension has been computer-optimized to produce the least trailing for 30-minute exposures. The tracking error to half an hour has been improved by about a factor of three. Negative numbers indicate that the mount is running fast—it is leading the sky slightly rather than falling behind right from the start, as in the previous table. Eventually, the sky catches up, then pulls ahead rapidly:

Drive Bolt	Base	Trail length after							
		5m	**10m**	**15m**	**20m**	**25m**	**30m**	**35m**	**40m**
¼ x 20	11.388"	-1.0	-1.9	-2.4	-2.1	-0.8	1.8	6.0	12.0s
³⁄₁₆ x 32	7.117"	-1.1	-2.1	-2.7	-2.6	-1.6	0.6	4.4	9.9s
1mm	227.8mm	-1.0	-1.9	-2.4	-2.1	-0.9	1.6	5.6	11.5s

My barn door was built around a 32-pitch bolt to the dimensions shown in the second line of this table. Its total error in half an hour from geometric principles is equivalent to the trailing seen in a stationary camera whose shutter is left open for only 3.4 seconds! It first runs 2.7 seconds ahead of the sky, and then falls 0.6 seconds behind. In practice, vibration, misalignment, and the 20-second pauses between twists of the drive bolt are the dominant sources of trailing.

Further details and more complex designs may be found in Ballard's *A Handbook for Star Trackers* and in the pages of *Astronomy* and *Sky and Telescope* magazines (see for example *S&T,* February 1988 and April 1989).

If the design and construction of the ultimate barn door is your goal, have fun. I prefer my deceptively simple hand-cranked version. It fits in any camera bag where it is always handy. It sacrifices excess accuracy for convenience. Using it reminds me of sextants, slide rules, and the Foucault mirror test. All are tools of uncanny precision born of simple parts.

approximates the challenges of deep-sky photography through the main telescope. It's good training, and it can produce some spectacular wide-field views of very faint and distant subjects.

Piggyback photography with long lenses acquaints us with the Tao of Guiding. Guiding is, according to your temperament, the characteristic torture of the deep-sky astrophotographer or a classic portal to your Zen-being. With a properly working and reasonably well-aligned equatorial mount, nothing could be simpler than keeping a star locked to the faintly glowing cross hairs of a guiding eyepiece. But few activities are as mindless. That few trophies are as satisfying as a perfectly guided astrophoto is testament to just how mindless the activity is: your greatest enemy is inattention, because as soon as your attention wanders, the guidestar will make a break for the edge of the field. As long as you watch, the star will remain centered.

And yet . . . constant vigilance at the guiding eyepiece leads to its own small tragedies: you imagine peregrinations of the guide star, which you correct with altogether real movements of the telescope. The product of too constant vigilance at the eyepiece is a field of tiny squiggles drawn in starlight.

Looking into the eyepiece once or twice per minute should suffice. If it does not and you are constantly making declination corrections, then the polar alignment of your mount is "too imperfect." Contrary to the earnest beliefs of my third-grade teacher, who never made an astrophoto in her life, perfection in some cases is always a question of degree: no mount is ever aligned as well as it might be. If you are forever chasing your guide star or waiting for it to catch up in right ascension, then your telescope drive has excessive slack or periodic gear error or is running at the wrong rate—or something is seriously out of balance. If it cannot be adjusted, your only options are installing better drive gears, such as those offered by Ed Byers or Thomas Mathes, or adding a computer-equipped drive system that electronically compensates for periodic gear error, or using a fully automatic, CCD-based guiding system (which may have its own ideas about just which idiosyncrasies in your drive train may be forgiven and which punished bizarrely).

The better your telescope mount is aligned on the pole, the less onerous

Fig. 6.4. Orion's belt and sword, with Rigel and Barnard's Loop as well as a -3 meteor. 135mm lens, 30 minutes at F4, Fujicolor 800 HG Plus from Mescalero Sands, New Mexico. The Comet West photograph is by George Kelley, Jr., from a farm near Glade Spring, Virginia. 15 minutes, 28mm lens at F2.8, on Plus-X film pushed to 800 ASA.

is guiding and the less vulnerable your photographs are to the dangers guiding introduces.

The polar axles of some mounts come equipped with a telescopic bore sight to allow easy but still approximate alignment on the pole. With careful polar alignment, a dual-axis drive corrector, and a rigidly mounted piggyback camera, very beautiful and effective deep-sky photography becomes possible. Telephoto lenses of 200 to 500mm focal length become powerful astrographs. A rough guide to effective magnification on 35mm film is to divide the focal length of the piggybacked lens (in millimeters) by 50. Thus a 300mm telephoto offers an image magnification similar to small, 6x binoculars. Its photographic reach is far, far deeper. With a 300mm lens I can photograph anything I can see in 12- to 16-inch telescopes (and then some). A 300mm F4.5 telephoto deserves respect! It is really a 3-inch astrograph designed to fit your camera.

In the photograph of Rho Ophiuchi, Antares, and M4, my 300mm lens reached well below 15th magnitude (this is a one-hour exposure on commercially hypersensitized Technical Pan). This lens, an old Nikkor telephoto, works best at F5.6.

Images made with fast, premium telephotos ("sports lenses") made to be sharp when used wide open rival the performance of Schmidt cameras. Though terribly expensive, a 300mm F2.8, 400mm F3.5, or 500mm F4 costs half as much as custom Schmidts and is useful for other, terrestrial subjects, too. It may take years of practice to rationalize spending months of gross pay on a telephoto lens, but be careful—it can be done.

Photographic lenses rarely bring all wavelengths to a common focus the way the best astronomical objectives do. Film is particularly sensitive to violet light, which is often especially sloppily focused. Stars (especially hot, young, bright stars) become large blobs (like Elvis), no matter how perfectly you focus and guide.

"Minus violet" filters from Lumicon cut off all light shorter than 4400 Angstroms, effectively trimming violet light from incoming starlight and dramatically sharpening photographic images. (As a bonus, they also exclude a good bit

Fig. 6.5. The Cocoon Nebula and its dark trail to M39 in Cygnus. 30 minutes, F5.6. 300mm telephoto, minus-violet filter, hypered Tech Pan. Texas Star Party, 1993.

When you are certain that the guide star never left the cross hairs, when you know you made very few, very small guiding corrections, and yet your stars are trailed (or multiple) then one of two things has happened. Either differential flexure has reminded you that you need a more rigid piggyback mounting arrangement, or the guide telescope's mirror has "flopped" ever so slightly during the exposure. Most popular Schmidt-Cassegrain and Maksutov telescopes are focused by moving the primary mirror fore and aft in the tube. Whenever the primary mirror is mobile, some slight play is inevitable; a mirror that shifts even a little as the telescope turns with the sky is fatal to precise guiding. There are reasons refractors make the best guide telescopes, and simplicity of mechanical construction is one of them. Refractors stay aimed where you point them.

Alan McClure and Robert Little demonstrated the power of piggyback photography in the 1960s. Chuck Vaughn, Johnny Horne, and Akira Fujii are three master practitioners today.

Stage 4: Astrophotography through the telescope

Photography through the telescope is the graduate school of hard knocks. Some few subjects are relatively easy. Returning to them again and again offers encouragement during long seasons of trial and error. The earth-lit nightside of the Moon, the moons of Jupiter, the crescent phase of Venus, the very brightest nebulae and clusters, and wide double stars are all relatively simple targets. At prime focus, these subjects require only short exposures (hence no manual guiding), and they are easy to see through the camera's viewfinder. This makes composition and focusing easy. In the case of focusing, "easy" is a relative term that depends (what doesn't?) on your personal standards of excellence.

In all prime-focus astrophotography, the camera replaces the eyepiece at the first (or "prime") focus of the telescope's objective. A simple adapter mates the camera's lens mount to a 1.250- or 2.000-inch tube that slides into the eyepiece holder. Instead of bringing an eyepiece to focus, the telescope's focuser moves the camera in or out until the objective projects a razor-sharp image directly onto the film. The position of best focus is surprisingly hard to find because the tolerance for best focus is surprisingly stringent. With an F6 telescope, the film plane must be positioned within just 0.04mm (40 microns, eight red blood corpuscles laid end to end)! At F10, the depth of focus is a comparatively generous 0.11mm, and at F16 it is a positively lackadaisical 0.28mm.

Exposures for the easiest targets range from a fraction of a second to a couple of minutes. At prime focus for most telescopes, try ISO 400 film for about $\frac{1}{60}$ second to photograph crescent Venus; try $\frac{1}{15}$ second to photo-

Fig. 6.7. Antares, the globular cluster M4, and an immense cloud of bright and dark gas and dust. A one-hour exposure through a minus-violet filter, F5.6, 300mm telephoto. Hypered Technical Pan. 1993 Texas Star Party.

graph bright doubles like Albireo. Try ½ to 5 seconds to capture Saturn and its rings (but don't expect to capture any smaller details with prime-focus photography). Try 5 to 30 seconds to capture Jupiter's four bright moons. Mirror slap and the vibration of the opening shutter are real concerns. Fast prime-focus exposures of ⅟60 to 1 second suffer more; for exposures of half a second or longer your can use a "hat trick": open the shutter while holding your hat in front of the telescope. Move the hat for the duration of the exposure, then close the shutter (or return the hat).

Five to 30 seconds is a good start for photographing the earthlit nightside of the Moon (which is particularly beautiful on color film in a deep blue sky when the Moon is very young). With very fast film, one to five minutes will capture the brightest deep-sky objects (the Ring Nebula in Lyra, the Orion Nebula, globular cluster M13 in Hercules, Omega Centauri or 47 Tucanae way down south).

In spite of all expectations, I've never found the sunlit Moon to be a particularly easy prime-focus target. It is big and it is bright, but it is also different from every other subject in the night sky. While we aim to boost contrast everywhere else, the Moon seems to have far too much contrast to start with. Techniques that work for us everywhere else work against us on the Moon. Photographers who attempt to shoot the Moon would do well to remember what they are photographing: effectively a very tall, very dark mountain in the harshest possible sunlight, the cloudless day to end all cloudless days.

The "sunny F16 rule" long used by terrestrial photographers provides a good first guess. With most objects lit by a Sun as bright as Earth's, a good starting point for an exposure in seconds (at F16) is the reciprocal of the ISO value of your film. This would suggest $\frac{1}{125}$ at F16 for the Moon on T-Max 100. But the Moon is very dark, as dark as a clean blackboard, so increase this

Fig. 6.8. Clockwise from lower left: the Double Cluster, 6 minutes at F6 with a 5-inch refractor on hypered Tech Pan; the Horsehead and Flame Nebulae in Orion with the same film and telescope at F4.5, through a hydrogen-alpha filter from a suburban backyard, 30 minutes; M42 is a 5-minute exposure at F4.5 using the same telescope on Fuji 1600 ASA film; the solar eclipse was caught on May 30, 1984, in Friendship, North Carolina. On Kodachrome 25, $\frac{1}{500}$ second at prime focus of a 3.5-inch Questar.

exposure by about two stops, to ⅟30 at F16 with ISO 100 film. Use ⅟15 or longer for the interesting, shadow-laced terminator between lunar day and lunar night. Photographing the Moon the way the eye sees it, with its wide range of brilliant light and deep shade, is emphatically not an easy project. If you are trying Moon photography without a driven telescope, don't overlook the rule of 1000 advanced earlier in this chapter: with a 2000mm telescope and no clock drive, your exposure must be shorter than ½ second. Above all, choose a developer and film combination to keep contrast high, but within earthly bounds.

To record fine lunar detail, you must use that do-anything film, Kodak Technical Pan. And you must process it for more than the pictorial contrast Technidol developer provides. Jean Dragesco, *in High Resolution Astrophotography,* praises the developers Kodak HC-110 (dilution "B") and Rodinol. TechPan processed in either developer is very contrasty by ordinary standards but far less contrasty than when developed in D-19 as for deep-sky photography. For Moon work, try TechPan developed in HC-110 (9 minutes, 20 degrees Celsius) or Rodinol (diluted 1:50, 12 minutes at the same temperature) with very little agitation. Thus processed, its effective film speed is about ISO 150.

Fig. 6.9. Made near Glade Spring, Virginia, in George Kelley's rural back yard: 1 hour at F4.5 on hypered Technical Pan, 5-inch refractor, minus-violet filter, tracked by an ST4 CCD attached to a 3-inch F16 guide telescope. M65, M66, and NGC 3628 constitute the "Trio in Leo." I've digitally removed an interloping asteroid that would confuse those using this image for starhopping!

Telescopic exposures of nebulae, clusters, and other highlights of the deep sky require exposures from a few to several minutes, up to an hour or two (or three or four). Deep-sky photography at prime focus is a complex art requiring far more specialized techniques than can be enumerated and developed here. Consider telescopic deep-sky photography the final exam for the guiding lessons started with a piggyback camera: polar alignment must be as perfect as your patience allows; ex-

Fig. 6.10. This is what differential tube flexure looks like! A one-hour esposure of M31 from Mescalero Sands, New Mexico, becomes a case study rather than a wall hanging. 5-inch refractor at F4.5 with minus-violet filter on hypered Tech Pan.

cursions by your guide star must be caught and corrected immediately. Watching experts work, I would swear they know when the star is about to move! Films whose acquaintance you made doing long-exposure barn-door and piggyback photography must be met again, and intimately. Nothing can move so much as a micron between the guidescope and the main telescope. Specialized techniques and even more specialized equipment are de rigeur. Your notes can never be too detailed, and every little thing you do must be recorded and controlled before telescopic deep-sky photography begins to seem routine.

Daphne and Tony Hallas, Jack Newton, Bill Harris, Chris Schur, Kim Zussman, Jason Ware, and Martin Germano are only a few names that belong on the honor roll of prime-focus astrophotographers. Watch the amateur-oriented monthly magazines and join the hurly-burly of electronic bulletin boards dedicated to amateur astronomy to keep current on what films and techniques contemporary astrophotographers use.

Still more esoteric techniques may beckon: eyepiece projection for large-scale images of planets (Donald Parker and Jean Dragesco demonstrate the state of the art), narrow-band interference filters for details on the Sun (Wolfgang Lille), three-filter true-color photography of the deep sky (pioneered by professional astronomer David Malin, practiced by amateurs Tony and Daphne Hallas and Bill and Sally Fletcher).

CCD—Charge Coupled Device—cameras use a solid-state detector instead of a photographic emulsion to record faint light from the deepest sky and fine detail in short exposures of our neighboring worlds. Theirs is a new world of sensitivity and yet another realm of critical demands. Digital photography using electronic detectors has inspired widespread acceptance of computer methods in image enhancement that may be applied to other

Fig. 6.11. The Whirlpool Galaxy photographed from Glade Spring, Virginia: 45 minutes at F4.5 behind a 5-inch refractor. Minus violet filter, hypered TechPan, CCD autoguided.

kinds of general-interest and astronomical photography as well.

Brad Wallis and Robert Provin offer a thorough technical introduction to astronomical photography in *A Manual of Advanced Celestial Photography.* This is not an introduction in the sense of being basic but in the strict sense that it offers key insights and rigorous methods, which in turn allow readers to tackle 99 percent of the projects most of us will ever dream up.

On a less technical plane, Henry Paul's *Outer Space Photography for Amateurs* is a classic survey of the field. Although dated and now out of print, it is filled with examples and techniques of historic and perennial interest. Other more modern sources include *Astrophotography* by Barry Gordon, and *Astrophotography for the Amateur* by Michael Covington. High-magnification planetary and lunar photography is treated in Dobbins, Parker, and Capen's *Observing and Photographing the Solar System,* and in Jean Dragesco's *High Resolution Astrophotography.*

Special problems: Astrophotography under city skies

Techniques for capturing the faint light of the remote night sky are all too effective for capturing the light of the glowing city sky, also. For most 20th-Century observers, light tens of millions of years in transit is drowned in a rising tide of streetlights in the last few microseconds of its journey.

First, push your luck: find out what you can get away with. In my backyard, where the naked-eye limiting magnitude is usually about 4, I can expose hypered Technical Pan for up to three minutes at the F4.5 focus of my 5-inch refractor. Longer exposures turn the film black from the light scattered by the

50,000 light-crazed souls who surround me. Surprisingly, such brief exposures are long enough to produce gratifying images of the brightest nebulae and star clusters.

The glowing sky is the most extended object there is, and extended objects are very sensitive to the F-ratio of the optics used to photograph them. If skyglow is a problem, try longer focal lengths: for any given aperture, the stars will come through relatively unchanged while the glowing foreground sky will become less prominent the larger the F-number. By using my refractor at F6, I can often get away with six-minute exposures. Now and then I can even make a ten-minute exposure at F6—enough to record 15th- to 16th-magnitude stars (including faint asteroids, Pluto, and the moons of the outer planets).

CCDs are especially rewarding for city observers because their useful dynamic range is so great and because their images are so readily manipulated. Removing city light is only a little more difficult than simply "subtracting" it from the original image and then "stretching" the brightness range. Faint objects just above the detector's threshold are rendered in brilliant tones. In photographic terms, using a CCD and image-processing software is like using an extremely versatile and responsive polycontrast printing system.

When conventional photographs are digitized, either by being transferred to PhotoCDs or scanned using desktop equipment marketed for personal and office use, they acquire some of the advantages of electronic photography. They have far more pixels than even state-of-the-art CCD images and they are easily enhanced using exactly the same techniques. Naturally, digitized photographs do not benefit from the great quantum efficiency, wide linear response, and extraordinary number of gray tones available in images made with solid-state detectors.

If you elect neither to flee to the country nor to turn to electronics to escape city skies, Jack Marling's specialty supply house Lumicon offers help: sharp-cutoff hydrogen-alpha bandpass filters neatly sidestep city light. These broadband filters let more than 90 percent of the deep red light emitted by hydrogen nebulae at 6563 Angstroms through to your film while excluding virtually all light of shorter wavelengths. City light, composed mostly of mercury and sodium-vapor emission lines, is simply turned off. So is the faint (but sometimes troublesome) natural airglow near 5580 and 6300 Angstroms and virtually all auroral light. Despite the 90 percent transmission at the crucial hydrogen-alpha wavelength, exposures through Lumicon's bandpass filters need to be about twice as long as unfiltered exposures made under a naturally dark sky. This is because nebulae shine at other wavelengths besides

Fig. 6.12. The image of the photo at right, modified to resemble the visual impression from a superb dark-sky site.

Fig. 6.13. The California Nebula. 5-inch refractor, F4.5, hydrogen-alpha filter, hypered Tech Pan: 30 minutes from a well-lit suburban backyard.

hydrogen-alpha and this extra nebular light is also blocked by the filter.

I made the photograph of the Rosette Nebula on page 126 through skies so bright that I could easily read my notes by the city's skyglow. Lumicon's hydrogen-alpha pass filter and digital "stretching" have effectively made my backyard a dark-sky photographic site.

You can substitute a Wrattan 92 filter for Lumicon's special-purpose filter. The Wrattan 92's transmission at the hydrogen-alpha wavelength is not quite as great, nor is its cutoff quite as sharp. However, it may be easy to purchase locally, less expensive, and available in a variety of special sizes.

Only Kodak's Technical Pan 2415 film is usefully sensitive to the deep red hydrogen light (Tech Pan was originally produced for solar astronomers whose science seems to revolve around this wavelength; only later did this ultra-fine-grain film find its way into general photographic use and then come full circle into amateur astronomers' hands). For deep-sky astrophotography, inherently slow Tech Pan needs to be hypersensitized by first being desiccated and then soaked in hydrogen gas. Wallis and Provin's book provides details for doing this yourself, or you can buy gas-hypersensitized film straight from Lumicon.

The human eye is almost completely insensitive to this light at the frontier of the infrared. It is ironic that to photograph the dark skies we used to enjoy from our own backyards, city lights force us to turn a blind eye to heaven!

My exposures through the hydrogen-alpha filter can be more than ten times longer than exposures without the filter. Stars two magnitudes fainter shine through the glowing murk. Nebulae I always regarded as impossibly faint are easy targets. Deeper challenges for every season wait on every star chart. As a bonus, the narrow spectral range of the H-alpha bandpass filter

Fig. 6.14. The Rosette was photographed through city lights—30 minutes at F4.5 with one of the hydrogen-alpha filters shown at bottom. Tucson glows in the twilight as seen from the Catalina Mountains; Jupiter and Venus to the right, the winter stars overhead (1 minute at F3.5, 16mm fisheye, 1600 ASA Konica negative film).

allows photographic lenses to produce pinpoint star images. But beware: pinpoint stars reveal tiny guiding errors—astrophotography is always about such trade-offs, most of them humbling.

Separate focus trials for hydrogen-alpha photography are absolutely required—lenses bring it to focus at a point significantly different from the point at which they focus visible light. Even pure reflectors will need to be refocused to use the hydrogen-alpha window, since inserting the filter into their converging light paths moves their focal point an appreciable distance "out."

Color astrophotography from the city is more problematic. If your skies are truly bright—if 3rd- and 4th-magnitude stars are only dimly visible—then color photos of glittering starfields are probably not possible. Less seriously light-polluted skies hold out some hope. Filters that selectively block only the principal emission lines of streetlights assist color photography, but they are not nearly as effective as sharp cutoff filters for black and white photography.

Both Lumicon and Tim Giessler's Orion Telescope Center (and a few other specialty shops) offer several types of LPR ("light pollution rejection") filters for visual observing and for color photography.

Throughout this discursive tour of astrophotography, names mix freely with techniques and technology. Astrophotography remains a small niche where individual practitioners make important discoveries and innovations. Such local fame does not usually come with a defensive mantle of isolation. Many of the masters are easily accessible via the Internet and on commercial online services. Read the magazines, study the books, seek out the photographers whose work you admire. Watch what others do, and by all means go outside, learn and improve these techniques yourself.

7

Rainy Night Astronomy

Into every life some rain must fall, and always, it seems, when clear skies are most wanted. Even though cloudy weather may inspire you to say things that frighten the tourists, a lot of astronomy takes place on rainy nights. It may be cold comfort if you have traveled halfway around the world at great expense to see an eclipse and are treated instead to the bottoms of clouds, but except in such dramatic settings, foul-weather astronomy can be as interesting as the fair-weather kind. Cloudy skies give us time to step back, take a break, catch our breath. Choose your metaphor as you will, but bad weather for viewing offers the chance (and the motivation) to see our obsession in other contexts, to enjoy astronomy in other ways.

A lot of people give up amateur astronomy for reasons having nothing whatsoever to do with a waning passion for the sky. The demands of school, a job, a family intrude. The normal hours of our social strictures contrast harshly with the best hours of our chosen hobby, and the collision often proves fatal to our astronomical interests.

There are ways to fit an astronomical obsession into normal life. Solutions range from a cataclysmic rearrangement of life itself (quit your job, leave hearth and home, and move to Arizona, consequences be damned) to the jejune (if you have to work nights, take up solar observing; if you cannot escape city lights but once a year, put the deep sky on hold and learn the face of the Moon). Saccharine aphorisms about lemons and lemonade come to mind. The challenge is to find a solution somewhere between the obsessive and the prosaic.

Isolation discourages more amateur astronomers than do all issues of time and opportunity combined. Not only must our hobby be pursued in hours that are regarded by our fellows as odd to the point of perversity, but our interests and our enthusiasms themselves often seem pathologically arcane. Some of us do eschew human companionship in favor of starlight, but this is not a level of commitment with which every amateur is comfortable.

Astronomy or a normal life? It is too easy to believe this is a choice that must be made, and it is an especially tempting thought at half past three in the morning when the temperature is zero, ice covers everything metallic,

Fig. 7.1. A dismal sight: the partially eclipsed Sun through thickening clouds. Iridescent clouds eventually gave way to rain.

and everyone whose face you remember is sound asleep and warm. Who wouldn't be susceptible to self-pity and to mortal misgivings? Even if you have a wonderful night of observing, even if you find some feeble prize you have sought for years, you can never expect your delight to be met by more than amused and bemused tolerance from most others.

The hobby can be as gregarious or as solitary as anyone could desire. You need not always observe alone, even if you generally prefer to. Do some sidewalk astronomy (set up a telescope on a street corner and show the night to the unlikeliest of passersby, following John Dobson's example) or provide telescopes and expertise for local parks and schools. Both projects connect you to your fellows and to the night.

If you need a hit of astronomical companionship only once a month or so, try an astronomy club. In America, most are affiliated with the Astronomical

League and can be found easily by contacting that organization or just by asking around at a library, college, or planetarium.

If your need for companionship yields to cybernetic solutions, join one of the electronic astronomy clubs on all the major online services. America Online, CompuServe, and GEnie all offer rich electronic pubs where amateur and professional astronomers hang out. (This book began on the CompuServe AstroForum in e-mail between its authors about improvements to a 10-inch telescope.) The Internet's usenet group sci.astro.amateur is a rough-and-tumble bar by comparison. Microsoft's network is just spooling up as I write this; plans for an amateur astronomy section there show great promise.

Dedicated bulletin boards, commercial forums, home pages on the WWW, and listservers are accessible everywhere telephone lines run. Taken together, online services make a market in used equipment, offer encyclopedic reference materials, provide archives of valuable data and software, and frequently facilitate an over-the-shoulder view of science in action and conversational access to expert and novice practitioners alike.

Finding astronomical kith and kin is one thing—people who share your interests are as close as your club or your modem. Involving domestic partners, children, and your astronomically innocent friends is something else. You may be passionate about glimpsing some nearly invisible something hidden deep in the night, but chances are that even your dearest friends' patience will wane long before your enthusiasm does. To the uninitiated, one look at the Moon is very much like another. Jupiter has a few cloud belts today, just as it did last week, just as it doubtless will next week.[1] "We saw that last time, Dad, why should we look at it again?" An astronomical passion that can outlive the harshest cold front may succumb to colder shoulders. Or, what may be worse, it may strain and outlive marriages and custodial relationships, too. You must scheme and cajole, plot and gently shepherd according to your style. If astronomy itself will not suffice, find ways to combine your passion with theirs.

An astronomical bent is a wonderful reason to travel. Solar eclipses have been my excuse to visit latitudes and cultures I would never have known otherwise. I admit that my appreciation for the North African Sahel is filtered through an Anglo-American lens, but I do know the way the keepers of the Saharan flocks move like cloud shadows over their cinnamon-colored pastures, how the first rain of the season breaks like an ocean wave on the streets of Dakar, and how a hand-bored flute sounds when played in the Saharan dusk.

---

[1] That some of us have seen Jupiter with only one major cloud belt is news that will be received as proof that you have missed the point.

You have to go there to see how the Sun rises heatless as the Moon yet seems to ring like a bell in the height of the summer sky. I know these things because the shadow of the Moon passed over, and I was there to see it.

Likewise, when Comet Kohoutek promised to shine brightly during an annular eclipse in Costa Rica, I awoke to the screams of howler monkeys every morning, and saw the Pacific surf flash like a liquid aurora at night. What little I know of snorkeling, I learned in Costa Rica. The Christmas celebration in the tiny town of Liberia may have had nothing to do with astronomy, but it had something to do with me.

If your partner, significant other, children, and friends cannot imagine being chilled to the bone for a glimpse of the Horsehead Nebula in the clear air behind an arctic front, perhaps the sheer panache of traveling halfway around the world will be more to their tastes. If you get to see a total solar eclipse into the bargain, so much the better. Who knows? With that kind of introduction, maybe some of your enthusiasm will prove contagious after all.

After following a 1973 eclipse to the Sahara desert, Joel Harris founded Twilight Tours to organize trips to other eclipse tracks. Cruise operators and travel agents vie for your traveling dollars before every solar eclipse, when bright comets beckon, whenever any event with astronomical cachet draws nigh. Look for their advertisements in popular science magazines a year or year and a half in advance. Some have more expertise than others with the specialized arcana of eclipse trekking (the logistics of moving bulky, expensive, delicate gear on and off airplanes, buses, boats and burros, of passing untaxed through the domains of often-skeptical customs agents, etc.). Ask around. And book early—there aren't that many seats on any specialized tour, and by the time the popular press nominates your eclipse as the next "event of the century" it may be too late. If you elect to go on your own, be flexible, and good luck!

For details about upcoming solar eclipses and for hard-won advice on how to catch them, see Joel Harris's and Rich Talcott's *Chasing the Shadow.*

The trip that made Joel a shadow-hound and introduced us both to the Sahara was organized by Massachusetts-based Earthwatch. Earthwatch specializes in bringing scientists and interested amateurs in many fields together to do field research. Their expeditions cover the globe geographically and tile the plain intellectually. Their astronomical adventures have included eclipse expeditions, archeoastronomical excavations, photometric stints at observatories, and surveys of prehistoric monuments to confirm or refute their astronomical significance. Other Earthwatch-supported expeditions range from zoology to archeology and back again.

Astronomical travel needn't imply expeditionary adventure in hot pursuit of solar eclipses and great comets. It need not involve overseas (or even overnight) travel. National Parks and public lands do more than pre-serve interesting landforms and pristine terrain. Almost as an accident, they preserve the starry night as well. Go into the darkness over the Great Smokies, above Yellowstone, into the night that fills and overflows the Grand Canyon. If the stars are your destination, such earthly treasures can be your companions'. In many places, a parking turnout will do for stargaz-ing. There's no need to worry with back-country permits and wilderness folderol on every exploit, let alone thousand-dollar airline tickets, visas, and immunizations.

Sparse population and frequent clear weather make the American south-west a paradise for astronomers, both professional and amateur. An embar-rassment of national parks and monuments makes it attractive on other grounds as well. Star Hill Inn of Sapello, New Mexico, offers an astronomers' retreat in the southern Rockies; the Prude Ranch of west Texas hosts the Texas Star Party every spring and is open as a guest ranch throughout the year. State Parks, National Forests, and other federal lands make up our communal backyard, where stars shine as they did for your grandparents, long before the age of streetlights.

Kitt Peak National Observatory near Tucson, Lowell Observatory near Flagstaff, the Very Large Array west of Socorro, New Mexico, and McDonald Observatory near Fort Davis, Texas, all offer tours. Most occasionally offer public viewing using world-class instruments or world-class sites. Other ob-servatories throughout the world offer tours and similar public access. Ask.

Party time!

Far down the Florida Keys in February, the Winter Star Party kicks off a year of star parties. Almost every weekend when the Moon is not full, some-where there is a star party to punctuate the astronomical calendar. Many are recognized for particular strengths. The Winter Star Party is noted for steady seeing. In late spring, the Texas Star Party is famed for dark skies.

In summer, Riverside (in California) and Stellafane (in Vermont, the grand-daddy of all star parties) showcase optical and mechanical innovation. A host of other gatherings enliven other weekends, some specialized, some of gen-eral interest. Observers arrange rendezvous in Nebraska, New Mexico, Montana, North Carolina and Oregon. CCD enthusiasts meet in Ontario. There are more astro-gatherings every year, and room for plenty more. All star par-ties promise better-than-usual skies and exceptional camaraderie, and they deliver both.

If you're a deep-sky observer, then for one week every spring, your family has a reunion at the Prude Ranch in the scrub-covered west Texas mountains.

At least once, you should go to the Texas Star Party and say hello to cousins whose names you don't yet know.

Events of the century

Beware the "event-of-the-century" phenomenon. I do not know how to explain this without seeming either cynical or naive, but there is a tendency among many travel agents and the popular press to misunderstand or to misconstrue the rarity of certain astronomical events. Total solar eclipses worth chasing come along every year or two. Pick one and meet the Moon's shadow where your interests and celestial geometry come together. Just remember that "Your last chance to see a total eclipse" is, by and large, nothing more than advertising copy.

Modest meteor showers occur several times each year. Under moonless skies, the annual Geminid, Perseid, Orionid, and Leonid showers produce 30 to 60 meteors per hour. Truly spectacular "meteor storms" are largely unpredictable, with one noteworthy exception.

Throughout the 1700s and 1800s, November's Leonid meteors put on a spectacular show every 33 years. When no storm was observed in 1866, 1900, or 1933, astronomers blamed an encounter with Jupiter for disrupting the stream of meteoroids and spoiling this "Old Faithful" of meteor storms. Despite this official epitaph, the Leonids of 1966 put on a heart-stopping show, one that may be repeated in 1999.

Be ready, but don't hock the farm for a chance to see it. For the record, I have every intention of being under dark, clear skies the night of November 17–18, 1999. I will not go to India, make a point of being out west, or arrange passage to Arabia, but I will be outside. In 1966, my father got me out of bed at 4 a.m. We shivered in the cold for half an hour. Nothing was happening when we went back to bed, leaving a crystal sky filled with the winter constellations. An hour later the stars began to fall. I keep a journal entry written by my best astronomical friend. It describes the wonders I so narrowly missed. First one, and then another meteor. And then more. Everywhere he looked, George says, a meteor was sure to appear. As dawn began, tens or hundreds of meteors appeared every minute. On his way to work, he watched fireballs streak through the pale blue sky. At its peak, in the hours before dawn over the American west, the 1966 Leonids rained 150,000 meteors per hour!

Previous Leonid showers earned a spot in Shoshone winter counts: "1833—stars fell like snow." The 1966 Leonid storm found a place in pop music. (John Denver reported it a few years later in "Rocky Mountain High.") Observers on Kitt Peak gave the winter count a modern twist. The meteor shower was dizzying, they said, "like driving through heavy snow with your high beams on." As for myself—I slept through it.

Comets are the jokers in the cosmic deck. Great things are expected of many newfound comets, yet comets are notoriously unpredictable. They flare when far from the Sun, promising wonders. They gutter to near invisibility just when geometry says they should be blooming. Sometimes they brighten steadily, raising hopes, only to fade away to mediocrity. But sometimes they deliver.

Great comets of the 19th century and before seem altogether more glorious than comets of our time. Is there really a dearth of great comets, or did the darker skies of yesterday simply frame comets more dramatically? Does gaudy, fast-paced competition from TV detract from modern comet shows? 1957's Arend-Roland, 1965's Ikeya-Seki, 1970's Bennett, and 1976's West, and 1996's Hyakutake all put on excellent shows. Yet none, arguably, rivaled the Great Comets recorded by observers in previous centuries.

The collision in 1994 of Comet Shoemaker-Levy 9 with Jupiter lived up to its billing (and then some), but 1973's Comet Kohoutek did not. In 1986, Halley's Comet met astronomers' every expectation, but the public at large was disappointed. The comet's passage was less dramatic than in 1910. Its trajectory carried it far to the south at its brightest (thousands of observers followed it to Australia and South Africa). The biggest difference in Halley's Comet may have been political rather than physical. In 1910, America was a rural nation; by 1986, most of us had moved to the city where we leave the lights on. Halley's appearances are literally once-in-a-lifetime events. The comet sweeps through the inner solar system only once every 76 years. It's not every comet that can survive the anxious anticipation of two generations without disappointing the third!

Perhaps Comet Hale-Bopp, discovered in 1995 by two amateur astronomers and sure to be at its best in 1997, will really be the "comet of the century." While we prepare this book, excitement is rising. Hale-Bopp has already been compared to the Great Comet of 1811 by people who should know. Perhaps Hale-Bopp will show us a spectacle such as our forebears enjoyed: a pale star hanging in the evening sky with a tail like a lighthouse beam shining among the planets. Or maybe not.

Perspective! Knowing what to look for in the eyepiece informs our vision. Informed vision precludes bitter disappointment when the Whirlpool Galaxy fails to whirl before our eyes and allows it, and so many other sights, to dazzle us with subtle pyrotechnics if only we appreciate what we *can* see. Knowing what to expect from an "event of the century" is likewise a question of familiarity with what has happened before, with what might happen, and with what is actually happening now. An appreciation for what is truly rare and what is hyperbole immunizes you against both super-adrenalized spending and cynicism-inducing disappointment.

When convenient, blame the media

In 1924, Edwin E. Slosson, editor of the first science-writing syndicate in America, described his view of science journalism.

"The public that we are trying to reach is in the cultural stage when three-headed cows, Siamese twins and bearded ladies draw the crowds to the side shows." That is why, he explained, science is usually reported in short paragraphs ending in "-est." "The fastest or the slowest, the hottest or the coldest, the biggest or the smallest, and in any case the newest thing in the world" (Dorothy Nelkin, *Selling Science*).

In the popular press, the astronomical adjectives of choice are still "first," "last," "brightest," "closest," "biggest," and "farthest." They are conjoined when possible with intimations of planetary catastrophe or promises of limitless energy. Never mind that "last" seems to mean only "until next time" and that limitless energy might *be* a planetary catastrophe.

Casual familiarity with the specialized literature and a modicum of experience under the sky put you miles ahead of the newswriter at your local TV station. He rips the news from the wire and tells the talking head to gasp where indicated. His copy comes from an anonymous stringer who has never seen sunrise on the floor of Plato or watched the zodiacal light fade into the dawn. Subscribe to *Astronomy* or *Sky and Telescope* and haunt your library's periodicals shelves. For more technical information and for pointers into the hardcore literature of working science, *Science News* is an excellent weekly digest with plenty of astronomical flavor. All these sources provide accurate coverage of astronomical events without the obligatory veneer of breathless anticipation required before celestial happenings command space in the mainstream press.

Online services and astronomy clubs are excellent founts from which to bottle experience and perspective. In addition to commentary from experienced observers, online services offer near-real-time information from sources like the International Astronomical Union Circulars. Astronomical news in the electronic community is only hours—or minutes—removed from the professional community's best sources. In some cases, it's even ahead of the professional game, as when Atlanta amateur astronomers discovered a supernova in the heart of galaxy M51 and reported it online even before the IAU announced their discovery. When a new comet appears, or a supernova detonates in the heart of a nearby galaxy, word gets around. The actual text of the IAU Circulars is proprietary and is not supposed to be distributed

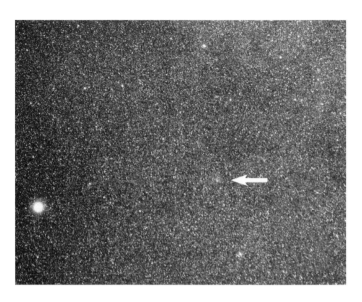

Fig. 7.2. Is this the next "comet of the century"? As this book goes to press in early 1996, hopes are rising for Hale-Bopp to put on a great show in 1997. But there's always another "event of the century," so beware spending time, money, and effort out of misplaced urgency.

verbatim beyond individuals and institutions who receive them by subscription from the Center for Astrophysics. The full text is available in many places online, though improperly so. Network firewalls are not always what they should be, and some sites post the Circulars intentionally. Go online and hunt these systems if you wish; in deference to the Center for Astrophysics, I don't feel comfortable offering further hints.

Impossible postcards

Flawed news and flawed features are causally related: in both cases, earnest reporters and sincere craftsmen simply do not know the sins they commit. In any rack of postcards, and in an embarrassing number of books, you can find the Full Moon glibly, impossibly, pasted into a sunset. Orion—every fiction writer's favorite constellation—turns up in every possible context, often months out of season. No less a magazine than *National Geographic* commits its share of astronomical blunders (Moons passing "overhead" when they could be no more than halfway up the sky; a photo of kids playing baseball captioned to indicate that they are doing so by the light of the midnight Sun even though a map only a few pages away clearly shows that their venue is far south of the Arctic Circle).

How many times in movies and in television advertisements have you seen a "reversed sunset" offered as a sunrise? The rising Sun, as seen from northern temperate latitudes, always moves to the right as it lifts from the horizon. A reversed sunset will move to the left unless the photographer thinks to flip the film as well as run it in reverse. It can't just be a matter of convenience and easy composition: even animated sunrises often show the

Sun moving the wrong way. When such gaffes catch your eye, congratulations! It means you're paying attention, and better connected to the world than most of your neighbors are.

Intellectual compartmentalization

Not only does astronomy tend to isolate us from one another, it tends to settle into a niche far removed from our other interests. This too is a rainy night concern.

I have a friend who doesn't really care a thing about astronomy. History is his thing. Whenever we talked, I felt compelled to set my astronomical interests aside. In a slow-motion capture dance (like Shoemaker-Levy 9 circling back toward Jupiter), our interests came together over the Missouri River of Thomas Jefferson's day.

We set out to read the million-word diaries of Captains Lewis and Clark, in "real time," as it were. We read, for instance their journal entries for May 29, 1805, on May 29, 1995. This deliberate pace gave us a chance to appreciate the scale of North America when first encountered by the young nation, gave us time to fit the ambitious reading into our working days, and guaranteed that we would have time to look into the inevitable mysterious detail. It kept our readings in step and forced upon us a sense of how the expedition got on day by working day.

So, on the partly cloudy night of January 14–15, 1805, with the temperature 10 below zero Fahrenheit, we found Meriwether Lewis trying to refine the position of Fort Mandan by gazing at the Moon. These sailors of the high plains referred to the Moon for the same reason sailors of the high seas did: to measure their longitude. Lewis intended to borrow a time-check from the sky by observing a total eclipse of the Moon (creative spelling is a hallmark of both captains' journals):

"Observed an Eclips of the Moon. I had no other glass to assist me in this observation but a small refracting telescope belonging to my sextant, which however was of considerable service, as it enabled me to define the edge of the moon's immage with much more precision than I could have done with the natural eye. The commencement of the eclips was obscured by clouds. . . ."

The next time there is a partial or a total lunar eclipse, you try fixing its various stages to better than a few minutes time. Now imagine doing so with marginal optics (the telescope from Lewis's sextant probably magnified about 4x) while wrapped in a buffalo robe on a frosty winter night in North Dakota.

Why was Lewis using only a part of his sextant rather than a larger telescope? What about that "spyglass" shown in statues of the leaders of the Corps of Discovery? Why did Lewis adopt a complex system of lunar sightings to fix position and time rather than follow Sir Alexander Mackenzie's successful example from a decade before?

In 1789, Mackenzie discovered the second largest river on the continent and followed his "River of Disappointment" to the Arctic Ocean, all the while hoping it might somehow find the Pacific and provide an easy route across the continent. On that voyage, Mackenzie roused his camp-mates to admire the midnight Sun from the banks of what is now the Mackenzie River. He returned to Montreal defeated. Five years later, still a decade ahead of Lewis and Clark, he navigated a labyrinth of mountain streams to make his way to the Pacific. To fix the time in order to determine his longitude with confidence, Mackenzie also used eclipses. But the eclipses he observed were of the moons of Jupiter; they provided his time standard in the sky.

With a small telescope, watch the fast-moving inner moons of Jupiter emerge from eclipse. With practice, couldn't you get this down to a few seconds of time? Timetables for the phenomena of Jupiter's four bright moons appear in both major astronomy magazines—just as they were listed in almanacs of the explorers' day. Plenty of cheap (even free!) computer software is available now to generate predictions as you need them to watch the distant moons of Jupiter. (Systematic observations similar to Mackenzie's allowed Ole Roemer to make the first accurate measurement of the speed of light in the 17th century—can you devise Roemer's test and repeat it? Any history of physics will provide the recipe, though few explain it very clearly from an observer-practitioner's point of view. In any event, it's more fun to work it out from scratch.)

Using a desktop planetarium, it's easy to recreate historic skies on virtually any computer. They can answer other questions that swirl around this rare misadventure in the generally brilliant Lewis and Clark Expedition. How many lunar eclipses could the young captain have observed before that winter on the high plains? Did Lewis have any idea what to watch for—how inherently vague were the contacts of the Moon with the Earth's shadow? He needed to time these to seconds and this was futile from the start. Where was Jupiter during the expedition? On what days, how often, could he have consulted the moons of Jupiter as did his westering predecessor? To do this, he needed better optics than he carried. Was Jefferson's refractor really too bulky to bring along? Was it ever discussed? How did a good refractor of that day compare to any modern refractor, however humble? For that matter, why does the symbol for the planet Mercury precede the date June 13, 1804, in Lewis's "Codex O"?

What does Clark mean when he notes on August 27, 1804, "This morning the Star Calld. the morning Star [was] much larger than common"? August 27 was a busy day in South Dakota: George Droulliard returned to camp to say that he could not find George Shannon or the horses Shannon was out looking for, and two more men were dispatched to find them. The sail was raised, the poles manned, and the Corps of Discovery began, for another day, its Sisyphusian push against the Missouri's current. The one death on the expedition, that of Sergeant Charles Floyd, occurred only a week before; his successor, newly promoted Sergeant Gass, was beginning his first full day of command. Even so, the appearance of the morning star commanded Captain Clark's attention and merited a note in the daily record. The flights of Apollo, the space shuttle, the Hubble Space Telescope notwithstanding—just try to tell me we are less alienated from the sky in this "modern age."

Sheldon Cohen teaches ancient Greek philosophy at the University of Tennessee. With the aid of Paul Burke and planetary ephemerides expert Jean Meeus, he reconstructed an observation by Aristotle involving a star in Gemini and the planet Jupiter. Computers and interdisciplinary expertise allowed Sheldon to combine his professional interests with his astronomical hobby to come up with a gratifying datum. "As far as I know," Sheldon writes, "December 4–5, 337 B.C. is the sole date for which we can specify Aristotle's activities." And what was Aristotle doing? He was stargazing!

The skies of history are yours to rediscover—if you cannot get away to indulge a historical interest by plying the wake of Lewis and Clark, or going to Athens to stand in Aristotle's footsteps, let the computer show you the skies over their heads instead. Mark Haney's shareware program *SkyGlobe* deserves special mention: it is a particularly easy to use digital planetarium for DOS-based PCs. Think of it as a versatile electronic planisphere. The program can be found in many electronic repositories, including the astronomical computing library of CompuServe's AstroForum. It's as useful and convenient for planning next week's observing sessions as it is for tracking down historic skies.

SkyGlobe, Dance of the Planets, Guide, MegaStar, SkyMap, TheSky, The Earth Centered Universe, and other electronic star atlases and planetariums too numerous to mention offer different strengths (and make different demands). More specialized programs allow you to plot upcoming eclipses (*Solar,* freeware by Matt Merrill) and historic ones (*Sun Tracker Pro,* a commercial product from Zephyr Services). Some let you hunt or identify asteroids with arc-second accuracy (*Guide* does this very well among its other deep-sky and planetary abilities, *Asteroid Pro* does this as well as possible, and very little else).

Scene synthesis programs (*VistaPro*) and mapping programs (*Mars Explorer*) let you take a walk on Mars or animate a flight over its volcanoes. One program (*TheSky*) even lets you control a very real observatory telescope and take images and data using it wherever you may be. There's no point in enumerating what's available in the cyber world, and no conceivable way to

Fig. 7.3. Craters this size on the Moon are barely visible; there must be tens of thousands of them. But Meteor Crater near Winslow, Arizona, is the only impact structure you can actually fall into.

do justice to all that software's strengths. Computers and computer software change very quickly. Check this month's magazines, or get online and look around. Silicon's skies are never cloudy.

When computers exhaust their numbers, traditional libraries remain: in *How the Shaman Stole the Moon,* amateur astronomer William H. Calvin clambers over kivas and rock ledges, under arches and over canyon rims to see how ancient stargazers of the southwestern United States measured and marked the changing skies. *Living the Sky,* by Ray Williamson, provides a broader view of astronomical traditions in native America.

Any professional astronomer will couch the story of creation in terms of the Big Bang, either to deny or affirm the account's broad strokes. Physicists and mathematicians are today's astronomer-priests. Other bands, other tribes, tell other tales. The Wolf Band of the Pawnee were the premier stargazers of the plains. If modern creation stories about quantum fluctuations and space-time singularities leave you cold, get out under the stars and try on the creation tales of other cultures.

Black God made the stars, say the Navajo, placing the star we call Polaris first and then carefully arranging Revolving Male, Revolving Female, Rabbit

Tracks (which we would call the stinger of Scorpius), First Big One, and Dilyehe (the Pleiades). Black God carried the stars in his pouch. Coyote watched while Black God went about this work. Black God took the stars out one by one to set them in their proper places. You know Coyote—he has no patience and quickly grew restless. When he could stand Black God's methodical pace no longer, Coyote snatched his pouch away and scattered the stars across the sky.[2] On a rainy night, learn others' stories; on a clear one, learn their constellations.

Messing about in BOTEs

Once upon a time, "calculator" meant "slide rule," and "computers" were people paid to work with tables of numbers and pads of paper. Engineers and scientists took pride in their ability to do casual calculations efficiently, to reach subtle and interesting results without fuss and bother. Their calculations were done on the backs of whatever envelopes fell to hand, hence their fond description as "back-of-the-envelope" (or BOTE) calculations.

On rainy nights or fair, BOTEs allow you to estimate the brightness of streetlights on the Moon, or the distance between the dust motes of pre-Perseid meteoroids as they stream toward Earth on an August night. When the universe is too big to contemplate, it is not too big to calculate. Just as starhopping provides a technology-free path through the starry night, BOTE calculations provide a computer-free peek at a semiquantitative universe.

Not only are BOTEs entertaining, they provide excellent demonstrations of the power of simple mathematics. If you know a junior astronomer who just can't stay interested in the "puzzle problems" offered in run-of-the-mill math texts, then here is a universe of alternative ones, each with fiery stars and spinning planets at its heart. I confess that I had no interest whatsoever in sines and cosines until an article by Isaac Asimov inspired me to use back-of-the-envelope trigonometry to see the moons of Jupiter as they appeared from one another, gliding through their own cold skies. That self-assigned project gave me the world of mathematics in all its variety and power and opportunity.

Water Rat in *The Wind in the Willows* is clearly a rodent after my own heart, for he tells Mole early on that "there is nothing, absolutely nothing, half so much worth doing as simply messing about in BOTEs."

Burnham's Celestial Handbook contains plenty of the grist needed for BOTE explorations (the mass and luminosity of other suns, for instance, the

---

[2]This version of "Coyote Scatters the Stars" is based on a rendering in *Archeoastronomy,* by Ron McCoy, 1992. It is V63, N2 of *Plateau,* published quarterly by the Museum of Northern Arizona, Flagstaff, Arizona.

distance between the components of double stars). Any introductory physics text will supply the simple formulae needed to get started along this path (the relation between light intensity and distance, or gravitational attraction and mass, orbital period and orbital radius). High-school physics and math offer plenty of power to bring the universe to life.

Computers and the BASIC programming language offer a way to power your BOTEs, to enlist the power of microcomputers to enhance your mathematical vision. If writing programs is not something you care to do, rest assured that plenty of us do enjoy cobbling up code to find impressive results, and we are always looking for an appreciative audience. Commercial programs, shareware, and freeware with astronomical themes are available in every level of sophistication for virtually every microcomputer on the market.

Give me darkness or give me death?

> *Science is a part of common culture, integrally tied to social practices, public policies and political affairs.* —Dorothy Nelkin, *Selling Science*

In the preface to the 1992 edition of her survey of science and media, Dorothy Nelkin had in mind "big" science, professional science, the science of controversial and expensive projects like space stations, super-colliders, and multinational field trips to Mars. But amateur science is no less (and possibly even more) closely tied to the structures and strictures of everyday life.

What follows is necessarily an American perspective. The issues will seem strange to Europeans, where doubtless there are others to take their place. These are live issues here in the "Excited States of America" and serve to demonstrate the connectedness of astronomy to other aspects of our lives.

Consider the 9mm in your eyepiece case. Is it a Nagler, or is it a Ruger? Optics or handgun? Does it make sense to go armed into that good night? Canadian Joe Yurchesyn watched a lively debate over that question on the CompuServe AstroForum and recounted some of its high points in the pages of *Sky and Telescope.* Amateur astronomers tend to observe alone, Joe notes, and "undesirable and antisocial activities are also carried out under dark, lonely conditions. As a result, amateur observing draws suspicion from the authorities, property owners, and passersby (both friendly and unfriendly)." How safe do you feel, parked at a remote wayside with several months' wages arrayed in glass around you? Does your trepidation center on a fear of personal harm or of fiscal loss? Is it appropriate to carry a deadly weapon to prevent either? Is it irresponsible not to? How about the accidents and mistakes that guns make lethal? Which concerns you more: hypothetical ne'er-do-wells roaming the remote countryside, hypothetical astronomers armed to the teeth, or a social milieu in which such questions need answers? Would you feel differently if you were of the opposite sex? Astronomy can force you to confront such questions; it isn't always the esoteric escape from gritty, real-world issues it may at first appear. Be prepared.

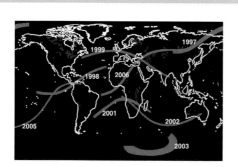

Against the madness of travel agents and ill-informed pressmen, here are maps and summaries of the next decade of eclipses. Pick one now and start making plans— nothing is more rewarding under cloudy skies than thinking far ahead. But beware "event-of-the-century" hype around eclipses: there's always another eclipse!

August 11, 1999

Total Solar Eclipses, 1996–2006

Date	Max. duration (m:ss)	Areas of visibility
1997, March 9	2:50	Siberia, Arctic Ocean
1998, February 26	4:08	Northernmost S. America, Caribbean Islands
1999, August 11	2:23	Cornwall, Central Europe, India. See map!
2001, June 21	4:56	South Atlantic, southern Africa
2002, December 4	2:04	Southern Africa, Indian Ocean, Australia
2003, November 23	1:57	Antarctica, Indian Ocean
2005, April 8	0:47	Total in south Pacific, annular in Panama, northern South America
2006, March 29	4:07	Northern Africa, Turkey, central Asia

Moon Shadows

Consider, as an example of BOTE computation, whether the bases of lunar landers left behind on the Moon cast shadows visible from Earth. A few key insights of BOTE technique are necessary. Primary among them is a trigonometric refrain. Repeat after me: "at small angles, the angle is equal to its tangent and the tangent is equal to the sine." Angles in BOTEs are best measured in radians, or units of $180/\pi$ degrees. One radian equals about 57 degrees. Close enough. A "small angle" in this context is one of less than about 1/10 radian, or 6 degrees.

From squinting at an Apollo snapshot, guess that the bases of the landers are about 20 feet wide and stand about 8 feet tall. (Precise numbers don't matter much in BOTEs—you can't be careless, but you can and must make some reasonable approximations along the way. Get the orders of magnitude right, get the first decimal place right, and if you have to keep more than three significant digits, you may be doing something wrong! Choose your approximations to keep the arithmetic simple and try not to do violence to your solution in the process.)

Any opaque object can cast a dark shadow to the distance at which it subtends the same angle as the illuminating object. From the Moon, the Sun appears essentially the same size it does from Earth, 1/2 degree, or 1/114 radian. From how far away will an object 20 feet wide subtend an angle of 1/114 radian? Recalling the first dictum about small angles and sines and tangents: from 114 x 20 feet = 2280 feet

Now we have the chance to refine our first, crude model: at that range, the height of the lander is only 40 percent the diameter of the Sun. So we note that the longest dark shadows are cast before the Sun has fully risen. And that suggests that the shadow might be slightly longer still (since the illuminating source is only part of a 1/2-degree-wide Sun). Increasing the length of the shadow by half again means that even less of the Sun can be covered by the 8-foot height, and that begins to suggest that the shadow will be projected across a deeply "twilit" landscape where it will be diffuse and hard to see. So a couple of effects are working against each other—the shadow could be longer than 2300 feet, but at the price of lower contrast. It seems that something only slightly longer than 2300 feet is likely. Say, less than 3000, more than 2000. Split the difference and settle on half a mile.

Half a mile at the distance of the Moon is .5/250,000 radians = 1/500,000 radian = 57/500,000 degree. An arc second (1/60 minute, itself 1/60 degree) is 1/3600 degree. So in seconds of arc, the longest shadow cast by a lander base is (57 x 3600)/500,000 seconds of arc, about

(55 x 4000)/500,000 seconds = 220,000/500,000 seconds. Call it half an arc second (and note that we just lengthened the shadow by several percent).

Under superb conditions, you might glimpse something half an arc second across on the Moon with any telescope bigger than about 8 to 10 inches (the diffraction limit in seconds of arc for any telescope is about 4.5/aperture in inches). But you could spy such a tiny object only if it were of the highest contrast. Linear features much smaller than the diffraction limit can be readily seen, but only if one dimension is significantly greater than that limit.

We need to know how much contrast the shadow of the lander offers. Consider: at most, we have a shadow 20 feet wide at its base tapering to nothing half a mile away. It covers roughly half of a rectangle 20 x 2500 feet or 25,000 square feet. The "blur circle" of a telescope resolving half an arc second at the Moon is a circle 2500 feet in diameter. It contains π x 1250 x 1250 square feet = about 3 x 1000 x 1500 = 4,500,000 square feet. Call it 5,000,000. So the shadow will appear no more than one part in 5,000,000/2,500 darker than its surroundings = 1 part in 5,000/2.5 = 1 part in 2000 darker. A photographic stop represents a change of 1/2, and this is a thousandth of that difference. We are close to the answer. We should reassess our assumptions, note the 10 percent here, the 5 percent there. Are any of those approximations in places that will propagate and make the final number very different? No. In plain language, you haven't got a chance.

8
What's Next?

I sometimes think there are more telescopes gathering dust than photons. If you've seen all the Messier objects, and followed the Moon's terminator from Picard to Grimaldi, and still find something . . . well, missing . . . don't worry. You've just hit the wall. Almost everyone experiences it, and while some push beyond, others simply call it quits. For the former, astronomy becomes a lifelong passion, but for the latter, their 'scopes are relegated to the garage, hauled out only for the astronomical event of the decade.

Astro fatigue comes in many forms: frustration with bad weather, a random or nonexistent viewing program, or simply being bored silly observing the same old objects. If this sounds familiar, read on! This is the chapter that can blow that layer of dust off your optics.

One of the easiest ways to get back on track is to join an organization dedicated to a branch of astronomy you love. If Jupiter is your thing, think about joining the Association for Lunar and Planetary Observers. Love the ever-changing world of sunspots? Contact the AAVSO Solar Division. If revisiting William Herschel's observations gives you a charge, the Astronomical League is just the ticket. No matter what your interest, there's a group for you!

The American Association of Variable Star Observers

If you think meaningful astronomical observations can only be done by professionals, think again. Since its founding in 1911, the American Association of Variable Star Observers (AAVSO) has provided amateurs with the chance to make observations that will benefit scientific research for years to come.

The AAVSO is a nonprofit organization created for those interested in observing stars which change in brightness—variable stars. With over 500 observers located around the world, the AAVSO has collected over 7 million observations, with between 240,000 and 265,000 observations sent in each year. Once received, they are processed and analyzed, with computer-plotted light curves created for all the variable stars in the AAVSO's program. The resultant data is published and then made available to professional researchers worldwide.

Professional astronomers seek AAVSO services for a variety of reasons, including real-time, up-to-date information on unusual stellar activity, assistance in scheduling variable star observing programs, and collaborative statistical analysis of stellar behavior using long-term AAVSO data. For example, a researcher may want to study a particular variable star during its "eruption," but has no way of knowing when an eruption will occur. Because of the millions of observations in their archives, the AAVSO has a good indication of when that event might occur. A network of amateur observers is then set up to keep a close watch on the target star; once the eruption is confirmed, the professionals are notified instantly.

Only a few hours of observing each month can make a considerable contribution, and any size telescope, or binoculars, is sufficient equipment. Using special report forms, your monthly observations are sent to the AAVSO Headquarters and at the end of the year, the names of observers and total observations are published in the Journal of the AAVSO.

According to the AAVSO, there is no minimum monthly observation quota—"even one observation is useful and significant." Just think, a researcher a hundred years from now may be relying on data from your nights under the stars.

AAVSO Nova Search Program. Memorize the sky? Are you nuts? Your friends may think you're crazy, but memorize the sky is just what you'll do if you wear the title Nova Hunter, for these galactic explorers search the stars for a new light where none has shone before. Under the auspices of the AAVSO Nova Search Committee, established in 1930, nova hunters search assigned areas of the sky, hoping to be the first person to witness a stellar outburst.

A systematic search for novae involves memorizing star patterns of the Milky Way. Many, like Kenneth Beckmann, Chairman of the Nova Search Committee, divide their binocular fields into small constellations of their own creation. In his book *Ghost Ships in the Night,* Beckmann describes a constellation of "geese returning south for the winter, and the starry outline of an M and M candy wrapper, or a sparkling Indian arrowhead." He writes, "Near the fifth magnitude star 73 Cygni, I have observed two cheerleaders successfully create a human pyramid. To the south, near the bright first magnitude star, Altair, I have witnessed the flight of a young eagle soaring about its mother, Aquila."

What equipment do you need for nova hunting? Nothing more than a pair of 7 x 50 binoculars and a good atlas. Here, the city observer actually has an edge. Because 8th-magnitude stars are the limit for visual nova hunting, light pollution actually helps mask out fainter stars. When you join the AAVSO and request information on nova hunting, you'll receive detailed information on conducting a systematic search.

According to the *Nova Hunter's Handbook,* "observers who systematically scan the heavens contribute knowledge about the distribution and frequency of novae in our galaxy. Once discovered, each nova adds a new chapter to the unfolding world of cataclysmic variables. Whether a novice or seasoned amateur, everyone who observes has an opportunity to build observing skills and perhaps, with a little luck, discover a nova."

AAVSO's American Sunspot Program. Sunspots have been studied since the beginning of recorded history. They have been thought to represent everything from dark clouds in the Sun's atmosphere to intra-Mercurial planets transiting the Sun. Heinrich Schwabe, a 19th-century amateur astronomer, discovered that the number of sunspots peaked at regular intervals. Soon afterwards, Rudolf Wolf, director of the Bern Observatory, began an effort to systematically record daily sunspot activity.

With the outbreak of World War II, obtaining solar activity data from Europe became a difficult undertaking. In 1944, astronomers Harlow Shapley and Donald Menzel suggested creating the American Sunspot Program, partly in response to the war problems, but also because it was felt that the United States needed an independent program. The observers for the program came from the ranks of the American Association of Variable Star Observers. Today, the program's members include observers from around the world.

If you like the excitement of tracking ever-changing sunspot groups, particularly when the Sun nears maximum, this program can be a lifelong project. The equipment needed is modest: a small scope, with good optics and a stable mount, equipped with a solar filter that fits over the front of the scope. With solar viewing, aperture size isn't nearly as important as the quality of the observer! In fact, the Wolf telescope, which is still in use, is only 80mm in diameter! For more information, contact:

American Association of Variable Star Observers
25 Birch Street
Cambridge, MA 02138-1205
617-354-0484
617-354-0665 fax
e-mail BITNET aavso@cfa8

or (for the Sunspot Program)
Peter Taylor
ptaylor@ngdc.noaa.gov

The Astronomical League

The Astronomical League (AL) is a federation of over 190 astronomical societies and individuals with over 11,600 members. If your astronomy club belongs to the AL, you're already a member. If not, you can join as a member-at-large, with a current yearly fee of $25. The league's purpose is to promote the science of astronomy; to this end it sponsors a variety of educational programs and encourages individual observers to explore the heavens through its observing award programs.

The AL holds a yearly convention featuring excellent guest speakers and programs. In 1995 the San Antonio Astronomical Association hosted the convention. Speakers included Frank Bash, Director of the McDonald Observatory, CCD image specialist Don Parker, and Deborah Byrd, *Astronomy* magazine author.

As a member, you'll receive the league's quarterly newsletter, the *Reflector,* along with a 10-percent discount on astronomy and related science books. In addition, member clubs can borrow films, slides, and videotapes. League members can also purchase, at minimal cost, a number of fine publications, including "A Guide to the Messier Objects," "Observe the Herschel Objects," and "Observe Meteors." The AL also publishes *Astro Notes,* which are concise descriptions of one aspect of amateur astronomy, such as how to use setting circles to find celestial objects, the different way time is measured in astronomy, a winter star-watching project to measure the effects of light pollution, and two techniques for accurate polar alignment.

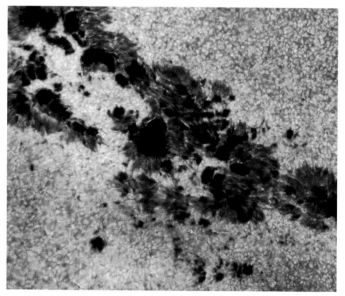

Fig. 8.1. Using only a small refractor on a stable mount, you can make meaningful observations when you join in the American Sunspot Program. This high-resolution photograph is courtesy National Optical Astronomy Observatories.

For the observer, the AL sponsors observer awards programs. In addition to issuing certificates for successful telescopic identification and logging of the Messier objects, double stars, and the Herschel 400, the league's John Wagoner has created three challenging observing programs for the binocular astronomer. These include the Messier, Deep Sky and Southern Skies Binocular Clubs. The requirements for each vary, and are detailed in John's publications. Generally, though, they require that you observe and log at least 50 (or more) specific objects. The Deep Sky program includes objects that take up where the Messier list left off, while the Southern Skies is a list for those living or traveling to the Southern Hemisphere. According to John, if you're going south on vacation or business, throw the list and a pair of binoculars in the suitcase, and try for this certificate. Combined, these three programs give an all-sky survey of deep-sky objects that are within range of small-aperture binoculars. And if you think your binoculars aren't big enough, John has observed most of the Messier list with a pair of 7 x 35s!

When you contact the league, they will send you a letter with the nearest member club, along with information on becoming a member-at-large. You will receive the *Reflector,* plus your choice of a subscription to one of five monthly astronomical publications. At this time, they include the Abrams Sky Calendar, *The Practical Observer, Griffith Observer, RASC Handbook* or *Star Date News.* If you're interested in joining the AL, contact their central office at:

Astronomical League
Science Service Building
1719 N Street N.W.
Washington, DC 20036
CompuServe 73357,1572

International Dark-Sky Association

When I was a kid, my family spent several days each summer on my uncle's Kansas farm, hand-cranking ice cream after supper and chasing fireflies at twilight. Later, lying on the still hot grass, we'd trace the weaving band of the Milky Way as it plowed across the black, black sky. I don't remember the last time I saw a firefly, but I do remember my last really dark sky: April 1994, Sentinel, Arizona.

A dark sky is a treasure. Not just for amateurs who eagerly wait each month's New Moon, but for scientists who turn their big glass further out into the universe. While the Hubble Space Telescope, along with other space-based observing platforms, is working miracles, there's still plenty of meaningful research being done right here on Earth. Unfortunately, though, our post-World War II generation has lit up the sky. Shopping malls, streetlights, car lots, security lights—they all contribute to a problem which has been created in our own lifetime—light pollution. When your neighbor's unshielded security light is on, a Full Moon shines not once a month, but every night.

While many yearn for the dark skies of childhood, others are doing something about it. In 1988, Dr. David Crawford formed a nonprofit organization called the International Dark-Sky Association (IDA), with the goals of effectively stopping the adverse environmental impact on dark skies by building an awareness of the issue of light pollution, and educating about the value and effectiveness of quality nighttime lighting.

Specific areas where the IDA is active include education, leaflets, economic information, lighting design examples, documentation of good versus bad lighting, and serving as a local resource to communities. The IDA currently has over 1500 members from 49 states and 56 countries. Of these, at least 135 are organizational members, including astronomy departments or observatories, amateur astronomy clubs and lighting companies.

With a current membership fee of $20, you will receive a newsletter and up-to-the-minute information as it becomes available. Should you have a local light-pollution problem, IDA will help with the solutions. And you may purchase IDA information sheets. They cover a wide range of issues and include "Recommendations for Effective Outdoor Lighting," "Star Watching Program," "Out to Get an Outdoor Lighting Ordinance" and "Lighting and Crime." To become a Dark Sky Advocate, contact:

Yes, you can make a difference in the battle over light pollution. Colorado observer David Kaufmann moved into a subdivision that had one streetlight for every three houses, with the streetlight on Dave's lot. Since the light was unshielded, he planned to try to control it through landscaping. However, in a conversation with a friend he learned that several cities in the United States are willing to work with homeowners in shielding lights.

Dave called the public utility and explained that he would like as little light as possible shining on his house and into his yard. Three weeks later they installed a fixture that retained the look of the original lantern-style light, but shines straight down. Dave plans to invite the rest of the neighbors over for a look through his scope, hoping this will prompt them to request new fixtures on other lights in the area.

Sometimes all it takes is a phone call.

International Dark-Sky Association
3545 N. Stewart
Tucson, AZ 85716

Association of Lunar and Planetary Observers

Who can forget their first view of Saturn, a truly "unearthly" planet with rings so crisp and sharp they appear surreal against the blackness of space? Or a night with seeing so steady you could swear you saw the rille in the floor of the Moon's Alpine Valley? The compelling sights in our own neighborhood of space are fortunately still visible even from light-polluted urban backyards.

The Association of Lunar and Planetary Observers was founded in 1947 to study our own Solar System bodies. It is an international group of students of the Sun, Moon, planets, asteroids, meteors and comets, whose goal is to promote the studies of these bodies using instruments and techniques generally available within the amateur community.

Sections have been established in many interest areas, including Mercury, Venus, Mars, Jupiter, and Saturn, as well as the Solar and Lunar Sections. Section Recorders gather observational material, encourage beginners, and contribute reports to the *ALPO Journal.* From time to time special projects may be initiated, such as the Lunar Transient Phenomena Section's observations, which were coordinated with the observations from the Clementine spacecraft in 1994.

What kind of projects do the ALPO Sections undertake? The Solar Section is interested in sunspot morphology, or how a sunspot changes over time. Observing the Sun in white light or by using a hydrogen-alpha filter, members have made over 7000 photographs and drawings that will aid professional astronomers in advancing knowledge of the Sun. The equipment needed is minimal, and the results rewarding. Do you enjoy viewing the Moon? If so, ALPO's Lunar Dome Survey needs you. If you have a large-aperture scope, the Saturn Section carries out visual and photographic observations of the ringed planet, including visual cartography, visual photometry, latitude measurements of belts and zones, and central meridian transits of visible details. One section requires no equipment but your unaided eyes—the Meteors Section.

Membership is currently $16 a year and includes a subscription to the Journal. For further information, contact:
Association of Lunar and Planetary Observers (A.L.P.O.)
Harry D. Jamieson
A.L.P.O. Membership Secretary
P.O. Box 16882
Memphis, TN 38186

Aurora Alert Hotline

David Huestis started the Aurora Alert Hotline in 1978 when, while working a second shift, he arrived home after midnight and found the skies blazing with a display of the northern lights. That year, several auroral displays were seen from his latitude of +42N, and when he mentioned these to friends their response was "you should have called me." From this grew the Aurora Alert Hotline, a network of phone callers, alerting others in the group to auroral displays.

Although some participants report data to interested organizations, the hotline is primarily for the aesthetic pleasure of watching Nature's color show. The four founding groups were Skyscrapers Inc., The Amateur Astronomical Society of Rhode Island, ATMs of Boston, and the Astronomical Society of Greater Hartford. Originally an organization of Northeast Observers, AAH now has observers from as far away as Yosemite, Washington State, and two Canadian provinces.

Because of the logistics of calling everyone in the group before a display is over, AAH is attempting to enlist individual clubs or groups in establishing their own hotlines. Since spectacular aurora displays are byproducts of solar activity, don't expect to see really outstanding events until around 2001 or 2002. For information on establishing your own hotline, contact:

David Huestis
Aurora Alert Hotline
25 Manley Drive
Pascoag, RI 02859

The American Meteor Society, Ltd.

The American Meteor Society was founded in 1911 for the purpose of visually following both meteor showers and meteors appearing at random. The AMS provides its members with brief observing directions, charts, and forms. All work can be done visually, without equipment, although the organization is also involved in CCD and radio meteor observations.

To qualify as an observer affiliate of AMS, you must go out on three clear, moonless nights, not earlier than 10 p.m. and make hourly counts of all the meteors you see, for at least three hours each night. You must record the count for each hour, giving hour and minute and starting and ending, along with notes on sky conditions. After the AMS has evaluated your observations it will determine if you qualify for affiliate membership. They believe that membership in the American Meteor Society will give you an opportunity to be trained in scientific observations, and in accuracy in reporting. For more information, contact:

The American Meteor Society, Ltd.
c/o David D. Meisel, Executive Director
Department of Physics and Astronomy
SUNY-College at Geneseo
1 College Circle
Geneseo, NY 14454-1484

International Amateur-Professional Photoelectric Photometry

An amateur astronomer with a small telescope in a backyard observatory can do real scientific research, obtaining data of professional quality by using the technique of photoelectric photometry. Photoelectric photometry makes it possible to measure the brightness of variable stars, for example, to a precision of 1 percent of 0.01 magnitude. In 1980 I.A.P.P.P. was formed to foster communication between professionals (who know of astrophysically significant problems that need solving) and amateurs (who can execute the projects). With over 1000 members representing over 50 countries, the organization is almost evenly divided between professionals and amateurs. For more information, contact:

Douglas Hall
Dyer Observatory
Vanderbilt University
Nashville, TN 37235
615-373-4897

Society for Amateur Scientists (SAS)

The Society for Amateur Scientists (SAS) is a nonprofit organization with a goal of helping anyone with a passion to do science take part in "the great scientific issues of our time." A unique collaboration between professionals and amateur scientists, SAS helps people develop their scientific skills to do significant research, even though they may not have the formal educational training generally required. SAS supports and encourages original research, with the goal of getting as many people involved in hands-on science as possible.

SAS projects are of a diverse nature and include research in most of the scientific disciplines, including botany, archaeology, seismology, and astronomy. In addition, the organization serves as an outreach program to education, developing projects for secondary and high school science classes.

Although established in 1994, SAS members have already done extensive measurements of cosmic ray rates during two eclipses and helped measure the ozone layer at the North Pole. Current astronomy projects include acquiring a library of astronomy software that will be made available on the Internet, and gaining access to professional instruments for amateur projects. Contact:

Society for Amateur Scientists
4951 D Clairemont Square, Ste 179
San Diego, CA 92117
World Wide Web http://www.thesphere.com/SAS/
comment@sas.org

Society of Amateur Radio Astronomers (SARA)

Imagine "listening" to Jupiter during the Great Comet Collision of 1994. Or tuning in as the Sun's nuclear engine revs up to the max! Or counting meteors not by eye, but by sound. This is the world of radio astronomy.

The job of the Society of Amateur Radio Astronomers (SARA) is to listen in on the universe. Formed in the early 1980s, SARA's members use modern, state-of-the-art, low-noise-receiving equipment for a variety of projects aimed at increasing scientific knowledge. Generally, the equipment consists of a good antenna system, a sensitive, stable low-noise receiver, and various output devices. The output may take the form of a strip-chart recorder, a voltmeter, or a data-logging computer.

Currently, SARA is encouraging radio observations of the galactic center and the Orion complex for anomalous pulses, which are believed to originate from these sources from time to time. Other members are involved in radio observation of the Sun, looking for flare activity at very low frequencies, while yet others are engaged in a supernova patrol. The American Meteor Society uses data supplied by the SARA members who log meteor infall counts, as radio detection of meteors is about ten times more effective than optical observations.

If you're new to radio astronomy, SARA members will be glad to help you get started. Contact:

Vincent Caracci
SARA
247 N. Linden Street
Massapequa, NY 11758

Sidewalk Astronomy

There's no organization to join, no dues to pay, no meetings to attend. Sidewalk astronomy is one of the ways all amateurs can help promote our hobby. All it takes is setting your telescope up in front of your house instead of behind it, and welcoming the neighbors to step right up and take a peek. Many amateurs conduct public star parties in conjunction with their own clubs, like Russell Sipe of the Orange County Astronomers, who holds public education programs at Palomar Mountain; some give free lectures to schools and church groups, while still others share in a far more informal setting.

Consider the story of Jim Marsh, a Missouri deep-sky observer.

When Jim fell in love with the stars, it became a lifelong affair. Born in a small town in rural Missouri, his first observing tools were his eyes and a 15-30 power rifle spotting scope. He didn't know formal constellations existed so he made up his own. In a town without library or bookstore, he learned the heavens by witnessing them first-hand.

The dawn of the Space Age found Jim in a large field behind Hannibal-LaGrange College, setting up a portable amateur radio rig just in time to hear the "beep-beep" heard round the Planet—Sputnik. Later, his view of Halley's Comet through the biology teacher's telescope was the beginning of a siren's song that's never been silenced. These days Jim observes through everything from a 4-inch Astroscan through a 25-inch Obsession, his passion the "faint fuzzies"—deep-sky objects.

Like John Chapman who seeded the orchards of the American Midwest, Jim travels the land, planting his love of galaxies and nebulae. He holds between 100 and 150 public programs a year and attends as many star parties as possible. During each, he remembers the warm welcome he received at his first Texas Star Party in 1988, and his resolve to make the newcomer feel at home. He calculates that over the years his scopes have "opened up the heavens to the eyes of over 100,000 people of all ages."

If you decide to follow in Jim's footsteps, congratulations!

Turning Pro

Perhaps the ultimate project, the ultimate way of "getting involved," is turning pro. If there is nothing in the wide world you would rather do than anything astronomical, then maybe it's worth considering. Astronomy brings a lot of talent into the sciences. It's flashy and "sexy" and far more rewarding to beginners' efforts than, say, particle physics is. Astronomy's practitioners, especially those who came to it from the amateur ranks, are genuinely, sometimes heroically, helpful to those who contemplate making the leap.

Youngsters have the inside track, because the road to professional astronomy leads through physics and serious mathematics. Begin early and stay with it. The standard advice is to major in physics (not astronomy) as an undergraduate and take as much math as you can. You may find a small, personal undergraduate environment is either nurturing or stultifying, but in

Fig. 8.2. The Milky Way arches over the upper telescope field at the Texas Star Party. The foreground observers and their telescopes are only a fraction of the 800 dedicated attendees at this annual event. Red lights protect night vision; the TSP's "dark police" confiscate white-light bulbs from guest rooms and tape light-proof insulation over all windows to help day sleepers and to prevent accidentally exposing observers to bright light!

either case aim for a high-powered graduate program, and always, always, look before you leap. Visit campuses, talk to students as well as to faculty. As your studies narrow, keep your interests broad.

People who hear astronomy's siren song later in life sometimes make excellent students and fine professionals, but more often than not they find supporting roles as educators, or as instrumentation specialists, engineers, and computer professionals attached to the science through universities and research institutes. They (we!) find astronomical gratification as supporting players and as "serious amateurs."

The number of people employed as full-time astronomers worldwide is not much different from the number of professionals in any major sport. Pursuing the dream of an astronomical career is a little like chasing "hoop dreams" with a Ph.D. Fortunately, many skills required for astronomy are valuable in other endeavors; there are far worse ambitions to steer an academic or professional career by.

Though slightly dated, Martin Cohen's *In Quest of Telescopes* is a frank and enthusiastic account of one professional's academic career from college days through several research appointments. The online world is ripe with up-to-the-minute real-world advice and information.

Spending your days and nights in observatories on remote mountaintops is an intoxicating idea, but this astronomer-as-forest-ranger image has little to do with the day-to-day work of the vast majority of professional astronomers. Most who succeed as pros are university professors; they teach physics to graduate and undergraduate students and pursue research in their chosen specialty whenever they possibly can. It's good work if you can get it, and if your temperament fits this reality.

Insecurity is the hallmark of many academic careers, and this one is no exception: one-year appointments wherever they may be, research niches that are often contingent on year-to-year funding, and continual scrutiny are the young professional's fare.

A science that amateurs pursue in the most beautiful of surroundings has a way of putting its pros through a long gauntlet of unfamiliar places and a multitude of trials. It's an odd combination of passions and skills that turns an amateur into a happy pro. If it's a leap you can't imagine not taking, study hard, and good luck!

Appendix: Suppliers and Publications

Astro Cards
P. O. Box 35
Natrona Heights, PA 15065

Astronomy Magazine
Kalmbach Publishing
21017 Crossroads Circle
P. O. Box 1612
Waukesha, WI 53187
800-533-6644

Atlas of the Moon
by Antonín Rükl
Kalmbach Publishing
ISBN 0-913135-17-8
800-533-6644

Beginner's Guide to the Sun
by Peter O. Taylor and Nancy
L. Hendrickson
Kalmbach Publishing
ISBN 0-913135-23-2
800-533-6644

Burnham's Celestial Handbook
by Robert Burnham
Available through Orion
Telescope Centers and
other telescope retailers

Chasing the Shadow:
An Observer's Guide to
Eclipses
by Joel Harris and Richard
Talcott
Kalmbach Publishing
ISBN 0-913135-21-6
800-533-6644

Earthwatch Expeditions, Inc.
319 Arlington St.
Watertown, MA 02172
617-926-8200
617-926-8532 (fax)

Exploring the Moon Through
Binoculars and Small
Telescopes
by Ernest Cherrington
ISBN 0-486-24491-1

Ghost Ships in the Night
by Kenneth Beckmann
P. O. Box 240
Lewiston, MI 49756

Guide
Project Pluto
Ridge Road, Box 1607
Bowdoinham, ME 04008
800-777-5886
pluto@genie.geis.com

KlassM Software
Mark Haney
P. O. Box 1067
Ann Arbor, MI 48106

Lumicon
2111 Research Drive 5A
Livermore, CA 94550
510-447-9570

Mag 6 Star Atlas
Available through Orion
Telescope Center and other
telescope product retailers

MegaStar for Windows
E.L.B. Software
8910 Willow Meadow Drive
Houston, TX 77031
713-541-9723

Orion Telescope Center
P. O. Box 1815-S
Santa Cruz, CA 95061-1815
800-447-1001 (order)
408-763-7030 (information)

Pocono Mountain Optics
800-569-4323 (order)
717-842-1500 (information)

A Portfolio of Lunar Drawings
by Harold Hill
Cambridge University Press
ISBN 0-521-38113-4

Prude Guest Ranch
P. O. Box 1431
Fort Davis, TX 79734
915-426-3203
915-426-3502 (fax)

Software Bisque
The Sky
012 Twelfth Street, Suite A
Golden, CO 80401
800-843-7599

Sky Atlas 2000
Available through Orion
Telescope Center and other
telescope product retailers

SkyMap
Chris Marriott
9 Severn Road, Culcheth
Cheshire WA3 5ED
United Kingdom
Internet
skymap@chrism.demon.co.uk
CompuServe 100113,1140

Sky and Telescope
Sky Publishing
P. O. Box 9111
Belmont, MA 02178

The Starry Messenger
P. O. Box 6552-G
Ithaca, NY 14851
201-992-6865
CompuServe 75020,3120
(Used astronomy equipment)

Telrad®
Steve Kufeld
1916 Woodland Drive
P. O. Box 6780
Pine Mountain, CA 93222
805-242-5421

Thousand Oaks Optical
Box 4813
Thousand Oaks, CA 91359
800-996-9111 (order)
805-491-3642

Roger W. Tuthill, Inc
11 Tanglewood Lane
Mountainside, NJ 07092
908-232-1786

Twilight Tours
3316 W. Chandler Blvd.
Burbank, CA 91505
818-841-8245

To contact the authors:
Nancy L. Hendrickson
CompuServe 73557,2602
Internet
73557.2602@compuserve.com

David Cortner
CompuServe 72550,322
Internet
72550.322@compuserve.com

Index